7th International Conference on Mechanical, Materials and Manufacturing (ICMMM 2020)

Held online due to COVID-19

IOP Conference Series: Materials Science and Engineering
Volume 999

Washington, DC, USA
25 - 27 September 2020

ISBN: 978-1-7138-3182-2
ISSN: 1757-8981

Printed from e-media with permission by:

Curran Associates, Inc.
57 Morehouse Lane
Red Hook, NY 12571

Some format issues inherent in the e-media version may also appear in this print version.

Additional copies of this publication are available from:

Curran Associates, Inc.
57 Morehouse Lane
Red Hook, NY 12571 USA
Phone: 845-758-0400
Fax: 845-758-2633
Email: curran@proceedings.com
Web: www.proceedings.com

TABLE OF CONTENTS

STRUCTURAL MECHANICS AND MATERIAL ANALYSIS

MATERIALS SCIENCE AND ENGINEERING

MECHANICAL DESIGN AND ADVANCED MANUFACTURING TECHNOLOGY

IOP Conf. Series: Materials Science and Engineering **999** (2020) 011001 doi:10.1088/1757-899X/999/1/011001

Preface

As the Chair of 2020 7th International Conference on Mechanical, Materials and Manufacturing (ICMMM 2020), I would like to say it is my honor to have been involved with this excellent and exciting conference.

Mechanical, materials and manufacturing are three disciplines that are all link directly. A conference that allows this latitude allows for research was success and the edge of knowledge can be pushed fully. This conference was no exception and I would like to thank the prestigious professors that supported the conference and presentations.

Many conferences claim to be international, very few are. We would have been in Maryland, just north of Washington DC, but circumstances have made our conference virtual. Nevertheless, we still have internationally spread presenters that justifies saying International.

Conferences need reviewers, and a thank you to all ours that supplied excellent feedback to maintain our standards. The level of presentation was as I expected, first class and not reduced by the virtual spect. These proceedings are to show the work and depth of research. We hope these will be the seminal work for larger research projects and cited as deserved. Finally, if you are reading and find interesting, submit a paper for next year, if you presented this year return and help us grow to deliver the support these disciplines deserved. Here is hoping we can all meet in person next year, as we plan to have the same venue as we would have this year. See you 25th – 27th September, 2021.

Warmest regards,

Professor Ian R. McAndrew
PhD FRAeS
ICMMM 2020 Conference Chair

IOP Conf. Series: Materials Science and Engineering **999** (2020) 011001 doi:10.1088/1757-899X/999/1/011001

Virtual Disclaimer

We would have been in Maryland, just north of Washington DC, but the travel restrictions because of COVID-19 have made our conference virtual. In order to meet the time of most participants, also not to affect the conference plans for next year, ICMMM 2020 was finally held during the period of September 24-26, 2020, as a fully virtual conference.

ICMMM 2020 consists of 3 keynote speeches and 2 parallel sessions. Over 30 participants attended online, they are from USA, Italy, China, Pakistan, India, Russia, Thailand, Colombia, Peru, etc. Each keynote speech lasted for 45min including 5min for Q&A, and for the parallel sessions, 15min each paper including 2-3min for Q&A. All participants were active, questions raised after every talk.

The whole conference was held online in ZOOM, which was technically supported by our conference affairs in Chengdu city, China. we also had test for each presenters in advance to ensure the success of delivery of the conference. For those who had internet problem, we also accepted a pre-recorded video presentation as an alternative of online presentation. Meanwhile the whole conference had been recorded for conference back-up only.

IOP Conf. Series: Materials Science and Engineering **999** (2020) 011001 doi:10.1088/1757-899X/999/1/011001

Conference Committees

Advisory Committee

Ramesh Agarwal (IEEE Fellow), Washington University in St. Louis, USA

Conference General Chair

Ian McAndrew, Capitol Technology University, USA

Conference Local Organizing Committee

Haijun Gong, Georgia Southern University, USA

Conference Program Chairs

Young Moon, Syracuse University, USA

Carl Moore, FAMU-FSU College of Engineering, USA

Conference Program Co-chairs

Michael E. Johnson, The Boeing Company, USA

Daniel Semere, KTH Royal Institute of Technology, Sweden

Przemysław Stach, College of Economics and Computer Science, Poland

Conference Technical Committees

Wang Yong, Nantong Vocational University, China

IOP Conf. Series: Materials Science and Engineering **999** (2020) 011001 doi:10.1088/1757-899X/999/1/011001

Peer review declaration

All papers published in this volume of *IOP Conference Series: Materials Science and Engineering* have been peer reviewed through processes administered by the Editors. Reviews were conducted by expert referees to the professional and scientific standards expected of a proceedings journal published by IOP Publishing.

- **Type of peer review:**

Double-blind

- **Conference submission management system:**

Iconf submission system

- **Number of submissions received:**

45 papers

- **Number of submissions sent for review:**

45 papers

- **Number of submissions accepted:**

17 papers

- **Acceptance Rate (Number of Submissions Accepted / Number of Submissions Received X 100):**

37.78%

- **Average number of reviews per paper:**

2 reviews

- **Total number of reviewers involved:**

24 reviewers

- **Any additional info on review process:**

1. Preliminary Review
2. The papers passed the first review will be eviewed again from the following aspects: Originality, Innovation, Technical Soundness, Applicability, Clarity of presentation and Relevance

Contact person for queries:

Ms. Weng P. L. Hwang (Conference Secretary), Email: icmmm@zhconf.ac.cn

Tel: +1-661-888-4278(USA); +86-28-83207566(China Branch)

IOP Conf. Series: Materials Science and Engineering **999** (2020) 012001 doi:10.1088/1757-899X/999/1/012001

Mechanical response of tensegrity dissipative devices incorporating shape memory alloys

N Singh[1], A Amendola[1], F Santos[2], G Benzoni[3], F Fraternali[1]

[1] Dept. of Civil Engineering, University of Salerno, Salerno,84084, Italy
[2] CERIS, Civil Engineering Department, Universidade NOVA de Lisboa, Portugal
[3] Dept. of Structural Engineering, University of California, San Diego, CA, USA
Email: snarinder@unisa.it; adaamendola1@unisa.it; fpas@fct.unl.pt; gbenzoni@eng.ucsd.edu; f.fraternali@unisa.it

Abstract. To optimize the seismic performance prescribed by modern structural codes, buildings and infrastructures must provide adequate safety for design level earthquake excitations, with limited levels of damage. This paper deals with the computational modelling of a bracing system with tensegrity architecture, which operates as a lightweight mechanical amplifier of longitudinal displacements in the transverse direction, efficiently limiting the inter-story drifts while dissipating energy. The proposed brace is based on a D-bar tensegrity structure with a rhomboidal shape comprising Shape-Memory Alloy (SMA) tendons. The SMA tendons can develop austenitic-martensitic (solid to solid) transformations, which enable them to amplify the signals into wide super elastic hysteresis, while subjected to mechanical cycles, comprising strains up to 6÷8%, with no residual deformations. The enhanced energy dissipation of the proposed SMA-D-bar (SMAD) braces are demonstrated through computational simulations of the response of braced frame to real earthquake events. The efficiency of the intended bracing to minimize the seismic impact of the served structure lays the foundation for the development of novel seismic energy dissipation systems integrating principles of tensegrity with superelasticity.

1. Introduction

Increasing attention are receiving mechanical metamaterials with unique dynamic properties not found in traditional materials [1]-[3]. Seismic metamaterials are unique; they either shield building including structures from seismic waves [4] or make use of novel seismic isolators based on lattice metamaterials, whose properties are mainly derived from their internal architecture, more than the nature of the component materials [5]-[6]. Metamaterials with nonlinear response can dynamically adjust their properties to the amplitude of traveling waves [7]. The category of nonlinear metamaterials featuring an internal tensegrity structure is especially significant since the response of such systems can be easily controlled by tuning internal and external prestress variables [8]. Dissipative bracing mechanisms provide effective resources in structures prone to extreme vibrations by passively dissipating the energy [9]. Such structures operate as nonlinear mechanisms and interact with the served structure, being capable of dissipating energy through, e.g., plastic response. In recent years, the use of displacement amplification dampers and superelastic Shape Memory Alloy (SMA) components within bracing systems has been extensively studied to account for re-centering capabilities [10]. The current research work deals with the use of an anti-seismic bracing device based on a tensegrity structure configured with SMA cables. The analyzed brace comprises a D-bar system, i.e., a planar tensegrity structure consisting of four bars shaped like a rhombus, which is balanced internally by inserting two perpendicular SMA cables (or strings).

Because of its geometric non-linearities, the SMAD brace examined here will significantly intensify the longitudinal displacement applied in the transverse direction, causing the transverse SMA cable to undergo a pronounced axial strain. Thus the structure can dissipate a large quantity of energy through its superelastic responses, without suffering significant deformations [11]. The SMAD brace's remarkable displacement amplification property is magnified when the device achieves a compact and tapered shape, which is especially convenient for designing non-invasive bracing systems. The geometry correlates with that of the scissor-jack damper described in ref.[12], with the difference that the SMAD brace replaces the viscous damper with SMA strings. It can be used profitably to shape novel seismic meta-materials inspired by tensegrity concepts [13]. After the installation of SMAD braces in benchmark building, a fragility analysis has been performed to illustrate their peculiar ability as anti-seismic protection gear.

2. Methodology

This study is focused on the development of a seismic bracing system with tensegrity architecture, which serves as a passive energy dissipation (PED) device, on combining noticeable displacement amplification properties with the superelastic response of SMA tendons. Various designs (as stated in Ref. [12]) were studied before adopting the scissor-Jack design by replacing the viscous damper with the SMA wires. A distinctive strategy has been incorporated in this structure and a system with D-Bars has been formed, as shown in Figure 1 (a), which shows the Scissor-Jack damper device equipped with SMA tendons. This design allows us to operate in a way that enables the element to be stretched while trying to move vertically.

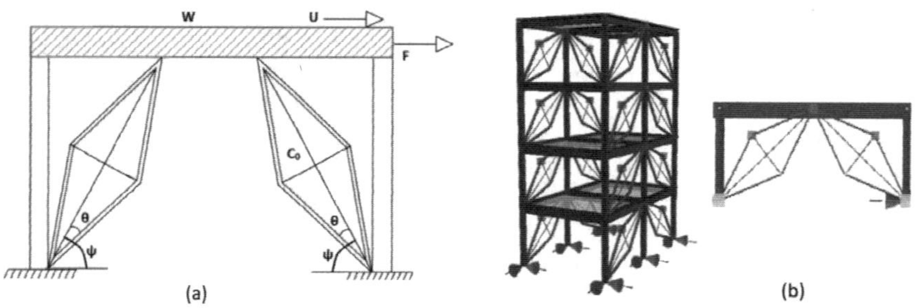

Figure 1. (a) Scissor-Jack Damper with SMAD brace, (b) Building model reinforced with C4T2 bracing devices.

Finally, the designed tensegrity-based device was installed on the benchmark structure and analyzed for its efficiency through fragility analysis, which is discussed in next section.

3. Numerical results

To computationally test the effects of the internal geometry on the mechanical response of the proposed unit, a small scale functional model of a SMAD brace was developed, which is hereafter referred to as C4T2 bracing (see Figure 1(b)). Both of the compressive struts consisted of two parallel 440 mm long pine wood bars with 35 upper 5 mm 2 cross-sections, joined at the mid-length of the bars between them. Hinges were mounted at the extremities to allow free rotation. NiTi superelastic strings (SMA wires) (d= 0.406 mm) were used to build the device tendons with a composition of 54.5-57 wt.-% Ni.

3.1 General description of the building

The research would concentrate on a 3-D model of the house (a four-story(4-story) steel building (Fig.1(b))), with a plan of 10.0 m by 6.0 m and altitude of 15.275 m. Throughout the x-direction, the bays are 5.0 m long, with a set of twice bays, and one separate bay in the y-direction. The building's lateral load-resistant mechanism consists of steel MRFs, with square tube sections and composite beams with broad girders on the flanges. Standard heights of floor to floor are 3.50 m. The bases of the columns are configured as fixed to the surface (at ground level). The floors are built of a composite structure consisting of wide-flange steel beams that work compositely with the floor slab, except for of the roof slab that is materialized by a reinforced concrete slab with a flat steel deck at its base. The floor structure includes diaphragm operation, and the horizontal plane is presumed to be rigid. It is presumed that the inertial effects of each point are conveyed uniformly to each perimeter of MRF by the floor diaphragm.

3.2 Ground motions for the structural analysis

To evaluate the seismic vulnerabilities and the damages caused by seismic actions, multiple strong ground movement scenarios were considered using the RSSIM program [15] relying on the non-stationary stochastic method [16] which takes into consideration the effects of finite faults. Using the earthquakes produced and incorporating different ground movements in two perpendicular directions, forty instances of seismic loading were created, ranging from 20.57o 94.64 s.

3.3 Seismic vulnerability assessment

The seismic vulnerability of the 4-story structure was assessed using a probabilistic method. Using a statistical approach [17], this method allows to consider the variability of the seismic activity and of multiple main FE model variables. In Table 1, the selected variables are considered to have normal distribution and are characterized.

Table 1. Probabilistic characterization of key parameters.

Variable	Units	Mean	St. Deviation
f_y (columns)	MPa	380.000	26.500
f_u (columns)	MPa	425.361	17.014
ε_u (columns)	%	23.421	1.405
f_y (beams)	MPa	300.000	21.000
f_u (beams)	MPa	438.841	17.554
ε_u (beams)	%	24.954	1.497
E	GPa	200	0.600
ν		0.300	0.009
γ (concrete)	kNm^{-3}	24.000	0.960

Successive incremental dynamic analyzes (IDAs) allow the identification of multiple damage states and the creation of fragility curves that can be used to estimate the probability of damage and the associated level of failure for provided peak ground accelerations (PGAs). To implement an IDA, a sequence of ground motion time-histories of increasing intensity are imposed on a nonlinear numerical model of the structure. In the present scenario, the PGA has been incrementally scaled (10 steps) from a low elastic response value, 0.1 g, up to 1.0 g, which ensures the failure in more than 50% of the FE model. To describe the damage levels, 80 IDA envelope curves were created by performing 640 nonlinear analyzes.

The assessment of the structural damage was performed using five damage state (DS) thresholds [18]-[19], which are DS0 = none, DS1 (slight) = 0.7 dy, DS2 (moderate) = dy, DS3 (extensive) = DS2 + 0.25 (du - dy), DS4 (collapse) = du. The description of the DS thresholds was made according to [20], using the generalized bilinear form of the curves of the IDA envelope.

The DS thresholds for an IDA envelope curve, together with its bilinearization are shown in given in Figure 2.

IOP Conf. Series: Materials Science and Engineering **999** (2020) 012001 doi:10.1088/1757-899X/999/1/012001

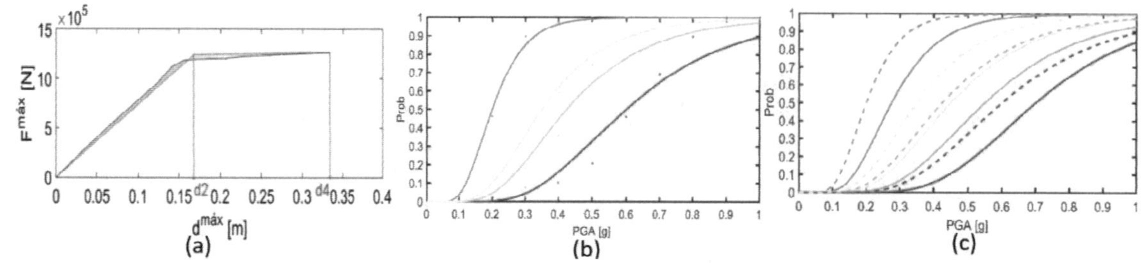

Figure 2. (a) Damage states thresholds for an envelope IDA envelope curve (b) Fragility curves of unbraced structure, (c) Fragility curves of braced structure

For DS threshold the associated fragility curve is defined as a function of the IDA displacement by the 50% probability of exceeding the respective threshold. Each fragility curve is defined by Equation 1, assuming it follows a standard log-normal cumulative distribution function φ.

$$P[DS_i/d] = \phi\left[\frac{1}{\beta_{DS_i}} ln\left(\frac{d}{DS_i}\right)\right] \tag{1}$$

Where d is the displacement of the IDA and βDSi is the standard deviation of the natural logarithm of DSi variable. The resulting global curves of fragility, for both the unbraced and C4T2 braced structures are given in Figures 2 (b) and 2 (c), respectively. To assess the seismic vulnerability of the structure, the probability of exceeding the DS thresholds for PGAs of 0.3 g (moderate earthquake ground motion) (Figure 3) and 0.65 g (strong earthquake ground motion) were computed. One can easily verify that the presence of proposed bracings in the structure decreases its seismic vulnerability.

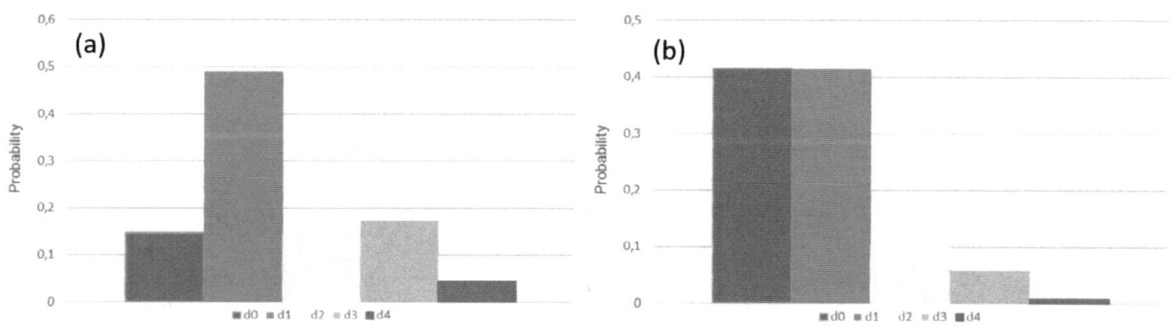

Figure 3. Probability of exceeding the DS thresholds for a PGA of 0.3g (a) unbraced structure (b) braced structures

4. Concluding remarks

The present paper has studied the seismic response of a tensegrity-based bracing with super elastic tendons. The given numerical simulations on a benchmark building model have shown that tensegrity inspired bracings can excel in structural seismic control. From the performed analysis one can indeed draw the following conclusions:

i. The compression efficiency of the proposed C4T2 bracing is associated with a geometrical advantage, combined with the use of tensile members that have been shown to exhibit large load to mass ratios.

ii. The proposed C4T2 tensegrity bracing acts as a mechanical amplifier for longitudinal displacements, increasing the level of deformations experienced by the transverse SMA tendons, and, hence, potentiating damping.

iii. In spite of its limitations, the proposed numerical model is able to reasonably apprehend the main features of the force-displacement diagrams shown by the proposed bracing.
iv. Based on the seismic vulnerability assessment it is possible to conclude that the proposed C4T2 bracings successfully mitigate the earthquake-induced damages, reducing the probability of collapse and that the proposed control approach shows great potential for seismic mitigation in civil engineering applications.

We address the experimental verification of the numerical results presented in this study to future work.

Acknowledgements
N.S., A.A. and F.F. acknowledge financial support from MIUR under the PRIN 2017 National Grant 'Multiscale Innovative Materials and Structures'(grant number 2017J4EAYB). The authors would like to thank Tiago Ricardo who worked on the subject covered by the present study during his MS Thesis work at the Universidade Nova de Lisboa, providing precious assistance with the numerical simulations presented in Section 3.

References
[1] Dan M B 2018 Sustainability 10.
[2] Christensen J, Kadic M, Kraft O and Wegener M 2015 *Mechanic research Communication* **5(3)** pp 453-462.
[3] Cummer S A Christensen J and Alu A 2016 *Nat Rev Materials* **1** 16001 .
[4] Kroedel S, Thome N and Daraio C 2015 *Extreme Mechanics Letters* **4** pp 111-117.
[5] Santos F, Benzoni G and Fraternali F Ingegneria Sismica / International Journal of Seismic Engineering **36(3)** pp 20-37.
[6] Fraternali F, Amendola A and Benzoni G 2018 *Ingegneria Sismica: International Journal of Earthquake Engineering* **4** pp 93-113.
[7] Hussein MI, Leamy M J and Ruzzene M 2014 *Applied Mechanics Reviews* **66(4)** pp 040802.
[8] Skelton RE and Oliveira MC 2010 *Tensegrity Systems*, Springer.
[9] Symans M D, Charney F A, Whittaker A S, Constantinou M C, Kircher C A, Johnson M W and McNamara R J 2008 *Journal of Structural Engineering* **134** pp 3-21.
[10] Menna C, Auricchio F and Asprone D 2015 *Shape memory alloy engineering. Boston: Butterworth- Heinemann,* p 369-403.
[11] Tanaka K, Kobayashi S and Sato Y 1986 *International Journal of Plasticity* **2** pp 59–72.
[12] Sigaher A N and Constantinouu M C 2003 *Earthquake Spectra* **19(1)** pp 133-158.
[13] Bertoldi K, Vitelli V, Christensen J and Van Hecke M 2017 *Nature Reviews Materials* **2** pp 17066 .
[14] Idelsohn S R and On˜ate E 1994 *International Journal for Numerical Methods in Engineering* **37** pp 3323–3341.
[15] Carvalho A, Zonno G, Franceschina G, Bile Serra J and Campos Costa A 2008 Soil Dy namics and Earthquake Engineering 28: 347-364.
[16] Carvalho A Campos C A and Sousa Oliveira C A 2004 13WCEE: *World Conference on Earthquake Engineering*, Vancouver B.C., Canada Paper no. 2194.
[17] Joint Committee on Structural Safety, JCSS Probabilistic Model Code, (online) available from URL: http://www.jcss.byg.dtu.dk (2001).
[18] Mouroux P and Brun B L 2006 *RISK-UE Project: An Advanced Approach to Earthquake Risk Scenarios With Application to Different European Towns Springer Netherlands*, Dordrecht p 479-508.
[19] Hazus-MH 2003 Multi-hazard loss estimation methodology: earthquake model, FEMA, http://www.fema.gov/hazus .
[20] Vargas Y F, Barbat A H, Pujades L G and Hurtado J E 2014 *Proceedings of the Institution of Civil Engineers Structures and Buildings* **167 (6)** pp 327–336.

IOP Conf. Series: Materials Science and Engineering **999** (2020) 012002 doi:10.1088/1757-899X/999/1/012002

On the fabrication and mechanical modelling microscale bistable tensegrity systems

Z Vangelatos[1], I Farina[2] A Micheletti[3], N Singh[4], C P Grigoropoulos[1], F Fraternali[4]

[1] Department of Mechanical Engineering, University of California, Berkeley, CA, USA.
[2] Department of Engineering, University of Naples Parthenope, Italy
[3] Department of Civil and Computer Science Engineering, University of Rome Tor Vergata, Italy
[4] Department of Civil Engineering, University of Salerno, Italy
Email: zacharias_vangelatos@berkeley.edu; micheletti@ing.uniroma2.it; snarinder@unisa.it; cgrigoro@berkeley.edu; f.fraternali@unisa.it

Abstract. We report about the analysis, design, and experimental testing of modular structures composed of bistable units derived from the classic triangular tensegrity prism. Tensegrity structures are pin-connected frameworks, composed by bars and cables, possessing internal mechanisms and self-stress states, and featuring a variety of structural responses depending on their prestress, edge connectivity, and geometry. When a tensegrity system has only one internal mechanism and one self-stress state, as in the triangular prism case, it is possible to associate to it a corresponding bistable unit, by replacing all cables with bars and changing their edge-lengths slightly. After presenting experimental results of compression tests carried out on microscale specimens fabricated through multiphoton lithography, we compare them with the numerical predictions obtained by our computational model.

1. Introduction

In recent times, architected metamaterials, structural systems whose physical behavior depends on the way they are designed rather than from the bulk properties of the constituent material, are attracting more and more attention from the scientific community. Metamaterials are designed to obtain extraordinary mechanical properties, such as exceptional strength-to-weight and stiffness-to-weight ratios, frequency bandgaps, negative overall elastic moduli, negative mass density, auxeticity, and solitary wave propagation [1]-[13]. Bistability is one of the difficult-to-find and most desired features in current studies on architected materials [14]-[19]. Further, tensegrity systems, pin-connected prestressed frameworks composed by bars and cables, constitute a particular structural class in that their response to statical and dynamical actions include a rich variety of non-linear effects [20][21][10]-[13]. We here focus on the Bistability property of some tensegrity structures [21] to design and fabricate bistable latticed structures at the nanoscale with the help of unique multiphoton lithography [23].

2. Bistable lattices with tensegrity architecture

Tensegrity systems are characterized by the existence of a self-stress state, and they can also possess internal mechanisms, which are nodal displacements which cause null first- order elongation of the elements, excluding rigid-body displacements. Self-stress states and mechanisms belongs to the left

IOP Conf. Series: Materials Science and Engineering **999** (2020) 012002 doi:10.1088/1757-899X/999/1/012002

and right null spaces of the equilibrium operator, expresses the linear relationship between the axial forces in the elements and the external loads [24].

We consider here the subclass of tensegrity systems possessing a single internal mechanism and a single self-stress state, which means that the equilibrium operator is represented by a rank- deficient square matrix, with rank deficiency equal to one. Stable tensegrity systems with internal mechanisms satisfy the so-called prestress-stability condition, signifying that the self- stress state imparts first-order geometric stiffness to any internal mechanism, which are then referred to as first-order infinitesimal mechanisms [24].

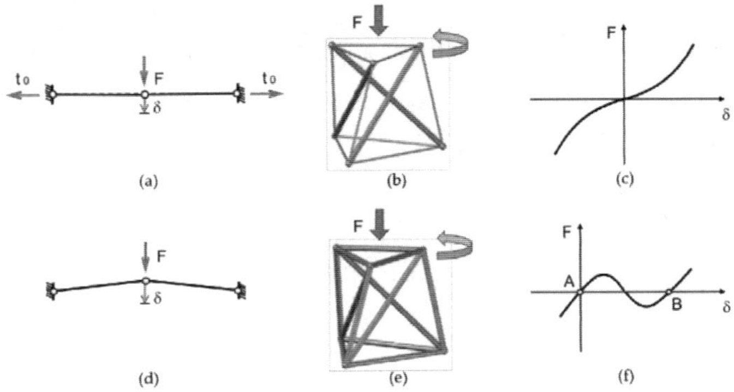

Figure 1. (a) Three-aligned-hinges system. (b) Triangular tensegrity prism. (c) Monostable load-vs-displacement plot. (d) Simple bistable system. (e) A bistable bar-framework corresponding to the tensegrity prism in (b). (f) Bistable load-vs-displacement plot. Curved green arrows in (b) and (e) indicate the twisting of the top base in the relative screw motion between bases.

Figure 1 illustrates the behavior of two prestress-stable systems, the simple three-aligned-hinges system Fig.1(a) [25], and the classical tensegrity prism (Fig. 1, b), when they are subjected to a load activating the mechanism. The typical load-vs-displacement curve is of stiffening type, and it can be approximated by a cubic with an inflexion point at the origin (Fig. 1, c). The slope at the origin is directly proportional to the level of self-stress t_0 in the elements. It can be verified that, by reversing the sign of prestress of a tensegrity system with the above-mentioned properties, the equilibrium configuration becomes unstable, and that two other stress-free stable equilibrium configurations arise (Fig. 1, d, e). The response to a load activating the internal mechanism is then of bistable type (Fig. 1, f) and it is associated to a double-well elastic energy. In the stable prestressed equilibrium configuration (triangular tensegrity prism), the top base is rotated with respect to the bottom base by an angle equal to $\theta_0 = \pi/6$, which is referred to as the *twist angle at equilibrium* (cf [20]). This value does not change when considering different circumscribed radii of the two bases, as shown in Fig.1(b). The first-order infinitesimal mechanisms are a relative screw motion (roto-translation) between bases, with the screw axis passing through the bases' centroids. The corresponding bistable unit is obtained by replacing cables with bars, and realizing the edge-lengths so as to have a twist angle slightly different from $\pi/6$. The structure is stress-free in this configuration, which we name as *primary stable configuration* (corresponding to point A in Fig. 1, f). When applying a vertical load to the top base, while having the bottom base fixed to the ground, the system snaps into the *secondary stable configuration* (corresponding to point B in Fig.1(f), through a relative screw motion between bases, analogous to the mechanism of the original tensegrity structure.

3. Fabrication by MPL and indentation experiments

We designed the modular assembly for experimentation as shown in Fig. 2, where three double units are assembled together side by side, a double unit being composed of two prisms

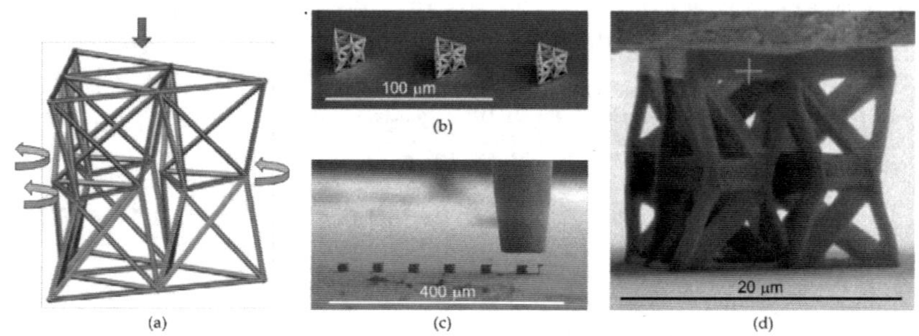

Figure 2. (a) Bistable assembly composed of six tensegrity prisms (one highlighted in red). (b, c) Fabricated samples. (d) View of a sample during the indentation experiment.

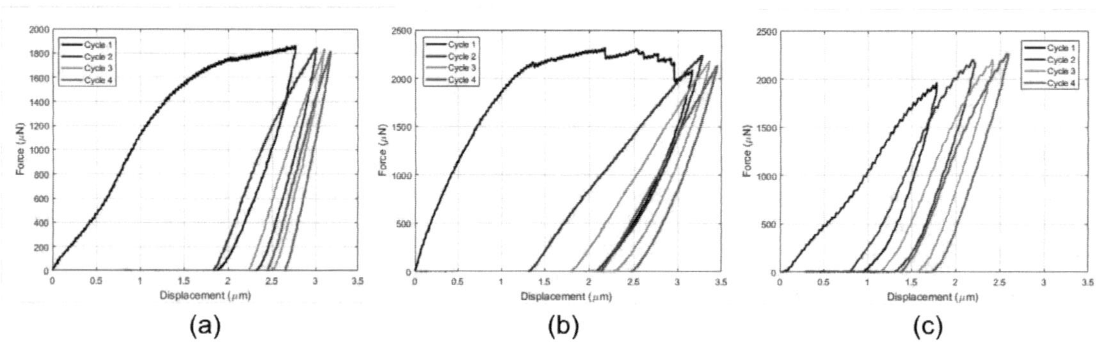

Figure 3. Force-vs-displacements plots for repeated loading-unloading cycles.

which is the mirror image of each other. The material for the specimens has been prepared by mixing organic-inorganic constituents in particular ratio as reported in ref. [22], such as: Zr-DMAEMA composed of zirconium prop oxide, (2-dimethylaminoethyl) methacrylate (DMAEMA) and ASTM type II deionized distilled water. After placing the mixture on a glass substrate in vacuum for 24 hours, specimen fabrication has been performed. Further details on the fabrication procedure are reported in [22]. The Young modulus of the fabricated material has been estimated with preliminary tests to be equal to E = 1.281 GPa. Indentation experiment have been performed by applying unilateral displacement-controlled loading-unloading cycles of increasing amplitude Fig. 2(c). Results of the testing are shown in Fig. 3. There is a marked softening behavior; the softening part of the curve leads to a secondary stable equilibrium configuration upon unloading; the secondary configuration is preserved during successive loading-unloading cycles; the curve presents a slight viscoelastic character. Moreover, it can be observed that, in association with the softening behavior, the three middle triangles undergo a twisting motion. In addition, the curves show that when the loads are close to the peak value, there can be microcracking events taking place termed as a sawtooth pattern in the curve.

4. Numerical experiments

We implemented a reduced-order model in a large-displacement regime based on the Stick &Spring approach (see, e.g. [26]-[28]). In the model, the nodal displacements are taken as Lagrangian parameters, while members can only extend or contract in a linearly elastic fashion, with stiffness constant ka, while they are rigid with respect to bending, shear, and torsion deformations. Additionally, the angular linearly elastic springs are placed at the nodes, with stiffness constant ks, to respond to changes in angle between the axes of pairs of members

IOP Conf. Series: Materials Science and Engineering **999** (2020) 012002 doi:10.1088/1757-899X/999/1/012002

which are not the part of a triangle.

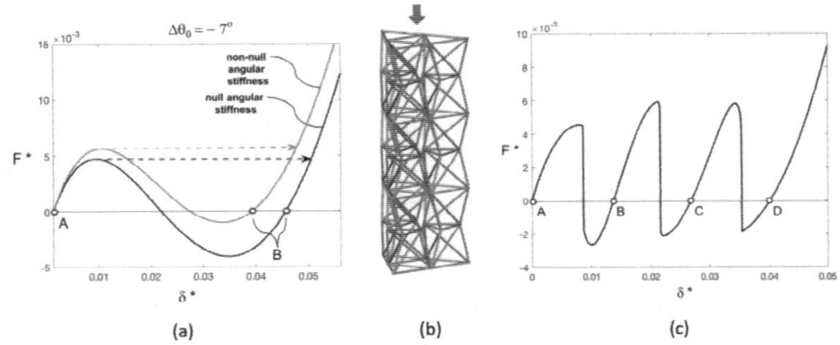

Figure 4. (a) Dimensionless force-vs-displacement plot obtained numerically for the bistable assembly, (b) Three-layer assembly. (c) Dimensionless force-vs-displacement plot of the three- layer assembly

Figure 4 shows the static response to a vertical downward load applied to the studied assembly. In the plot, the dimensionless load F^* is one third of the actual load divided by the stiffness constant of the shortest beam and by the beams' minimum diameter, while the dimensionless displacement δ is the vertical displacement divided by the height of the assembly. As to the geometric parameters in this simulation, we have that a/b = 0.7, a/h = 0.25, $\Delta\theta_0$ = 7degrees, where a and b are the circumscribed radii to the bottom and top bases of the unit, h is the height of the assembly, and $\Delta\theta_0 = \theta_0\ \pi/6$. The two curves differ in the value of the angular stiffness constant, this is null for the black curve, and it satisfies $k_s/(a^2k_a) = 0.00025$ for the color (grey) curve. The simulation is consistent with the fact that in a force-controlled loading-unloading experiment, the system would snap from a primary stable configuration (corresponding to point A in Fig. 4 (a)) to a secondary stable configuration (corresponding to points B in Fig. 4(a)), passing along a softening-type path before snapping. A behavior which is qualitatively similar to that observed in the indentation experiments, while the differences can be partly ascribed to the viscoelastic behavior of the photo resistive material. We then performed the same numerical experiment on a three-layer assembly obtained by superposing copies of the previously considered system on top of each other (Fig. 4, b). The three layers differ from one another in the stiffness of the angular springs, which are chosen with a 5 percent difference from one layer to another in order to obtain a sequential snapping of the layers. Geometric parameters are the same as in the previous assembly, and $k_s/(a^2k_a) = 0.00021$ for the middle layer. The resulting force-vs-displacement curve (Fig. 4, c) shows that there are multiple stable configurations (points A, B, C, D in Fig. 4, c), related to the bistable behavior of the individual layers.

5. Conclusions

In this research work, a bistable assembly composed of triangular tensegrity prisms has been designed, fabricated by MPL, and subjected to indentation testing.

The experimental force-vs-displacement curves highlight a softening response which brings the structure to a secondary stable equilibrium configuration. Such secondary configuration is maintained during successive loading-unloading cycles. In addition, a slight viscoelastic behavior is observed, together with occasional microcracking events.

A Stick & Spring reduced-order model is employed to simulate indentation experiment, confirming that the observed experimental behavior results from the bistable design of the constituent units.

A three-layer system is analyzed numerically to show that multi-stable lattices can be obtained.

Acknowledgements
This research is partially supported by the National Science Foundation (NSF) under the Scalable 475 Nanomanufacturing (SNM) Program, grant number 1449305. A.M. acknowledges financial support

from the Italian Ministry of Education, University, and Research (MIUR) under the PRIN 2017 National Grant '3D printing: a bridge to the future' (grant number 2017L7X3CS 004). N.S. and F.F. acknowledge financial support from MIUR under the PRIN 2017 National Grant 'Multiscale Innovative Materials and Structures' (grant number 2017J4EAYB).

References

[1] Liu Z, Zhang X, Mao Y, Zhu Y Y, Yang Z, Chan, C T and Sheng P 2000 *Science* **289** 1734–1736.
[2] Lu M H, Feng L and Chen Y F 2009 *Material Today* **12** pp 34-42.
[3] Maldovan M 2013 *Nature* **503** pp 209-217.
[4] Brunet T, Leng J and Mondain-Monva O 2013 *Science* **342** pp 323-324.
[5] Deng B, Mo C, Tournat V, Bertoldi K and Raney J R 2019 Physics Review Letter **123** pp 024101.
[6] Yildiz M E, Tran C A, Barchiesi E, Spagnuolo M, dell'Isola F and Hild F 2019 *State of the Art and Future Trends in Material Modeling* **100** p 485-505
[7] Meza L R, Das S and Greer J R 2014 *Science* **345** pp 1322–1326.
[8] Zheng X et al, 2014 *Science* **344** pp 6190.
[9] Phani A S, Hussein M I (eds) 2017 *Dynamics of Lattice Materials*, John Wiley & Sons, Ltd, Chichester, UK.
[10] Fraternali F, Senatore L and Daraio C 2012 *Journal of the Mechanics and Physics of Solids* **60** pp 1137-1144.
[11] Fraternali F, Carpentieri G, Amendola A, Skelton R E and Nesterenko V F 2014 *Applied Physics Letters* **105** pp 201903.
[12] Davini C, Micheletti A and Podio-Guidugli P 2016 *Meccanica* **51** pp 2763-2776.
[13] Micheletti A, Ruscica G and Fraternali F 2019 *Nonlinear Dynamics* **98** pp 2737-2753.
[14] Shan S, Kang SH, Raney J R, Wang P, Fang L, Candido F, Lewis J A and Bertoldi K 2015 *Advance Materials* **27** pp 4296-4301.
[15] Raney J R, Nadkarni N, Daraio C, Kochmann D M, Lewis J A and Bertoldi K 2016 *PNAS* **113** pp 9722-9727.
[16] Osama R B, Andr´e F and Chiara D 2017 *Proceedings of the National Academy of Sciences* **114** pp 4603-4606.
[17] Chen T., Bilal O R, Shea K and Daraio C 2018 *Proceedings of the National Academy of Sciences* **115** 5698- 5702.
[18] Deng B, Wang P, Tournat V and Bertoldi K 2019 *Journal of the Mechanics and Physics of Solids* pp 103661
[19] Jeong H Y, An S, Seo I C 2019 *Sci Rep* 9324.
[20] Oppenheim I J and Williams W O 2000 *Journal. Elasticity* **59,** pp. 51–65.
[21] Micheletti A. 2013 *Proceedings of Royal Society* A 469.
[22] Vangelatos Z, Micheletti A, P Grigoropoulos C and Fraternali F. 2020 *Nanomaterials* **10(4)** pp 652.
[23] Sakellari I, Kabouraki E, Gray D, Purlys V, Fotakis C, Pikulin A, Bityurin N, Vamvakaki M and Farsari M 2012 *ACS Nano* 6:2302-2311.
[24] Calladine C R and Pellegrino S 1991 *International Journal of Solid Structure* **27** pp 505-515.
[25] C R Calladine 1978 *International Journal of Solids and Structures* **14** pp 161-172.
[26] Favata A, Micheletti A and Podio-Guidugli P 2014 *International Journal of Engineering Sciences* **80** pp 4-20.
[27] Amendola A, Favata A and Micheletti A 2018 *Frontiers in Materials* **5** pp 1-22.
[28] Favata A, Micheletti A, Podio-Guidugli P and Pugno N M 2017 *Meccanica* **52** pp 1601-1624.

Mechanics of energy harvesters based on tensegrity solar facades

R Miranda[1], E Babilio[2], N Singh[1], D P Villamil[1], F Santos[3], F Fraternali[1]

[1] Department of Civil Engineering, University of Salerno, Salerno 84084, Italy
[2] Department of Structures for Engineering and Architecture (DiSt), University of Naples "Federico II", Naples 80134, Italy
[3] FCT, Civil Engineering Department, Universidade NOVA de Lisboa, 2829-516 Caparica, Portugal
Email: ramiranda@unisa.it; enrico.babilio@unina.it; snarinder@unisa.it; dianamaritzap@gmail.com; fpas@fct.unl.pt; f.fraternali@unisa.it

Abstract. This work is focused on the computational design of tensegrity shading systems of energy efficient buildings which mitigate air conditioning consumption and optimize the energy performance of the building. It is concerned with the design of active solar façade screens based on lightweight tensegrity units, which are easily integrated with energy harvesting piezoelectric cables. The unit cells of the analyzed screens are controlled by tensioning and releasing selected cables of the structure and are used to direct the shading panels towards the sun. A numerical procedure simulates the dynamics of the analyzed tensegrity façades, by considering the opening and closure motions of the screens, and the vibrations produced by the action of dynamic wind forces. The energy harvesting ability of the proposed façade is numerically estimated.

1 Introduction

The design of sustainable buildings naturally leads to analyze the fundamental elements of such buildings, which consist of mechanical ventilation systems, protective shading systems, and, eventually, home automation systems and renewable energy harvesting devices. Reduction of energy consumption and carbon dioxide emissions through intelligent and sustainable systems is one of the key objectives of EU's Horizon 2020 program "SC3" [1]. The demand for energy savings and the excessive CO_2 emissions generated by buildings has called for the use of new "interactive building envelopes", defined as systems that interacts with the variations of the external climatic conditions by means of automatic control devices, optimizing the environmental performance of the internal microclimate. This paper presents a mechanical study of dynamic sunscreens with tensegrity systems for the re-designing of the Al Bahar Towers facade in Abu Dhabi [2]-[4]. We enrich the kinematical study presented in [2] by providing an analytical procedure for the implementation of the system and the dynamics under wind forces. We employ such a procedure for the simulation of the motion exhibited by origami sunscreens under fluctuating (positive/negative) wind forces. The wind-induced energy stored in the activation strings is investigated with the aim of exploring the potential of the analyzed sunscreens as wind power harvesters. Their energy harvesting capacity is compared with that of microeolic turbines available in the market [5]-[7]. The results show that tensegrity solar façades as wind power harvesters are highly feasible, and may lead to design of cost-effective shading systems, which are energetically autonomous and do not require power from the electric grid for their operation.

2 WTABS Model

In the present study a tensegrity module, addressed as 'wind tensegrity Al Bahar screen' (WTABS), is analyzed in order to evaluate its ability in harvesting wind energy [8]. In order to harvest mechanical energy, the TABS module is enhanced with a number of D-bars equipped with strings wrapped by PVDF piezoelectric films as shown in Figure 1, which describes the full WTABS model. The actuation mechanism is generated by a piston Rolaram® R2501190, with design characteristics of dynamic loading equal to 294 kN and a maximum stroke of 3500 mm [9], which is located on the highest point of the WTABS structure. Such a piston stretches and releases the perimetric strings, thus activating the dynamic motion. A telescopic sliding constraint positioned behind the structure adjusts the scrolling along the z axis, while two linear springs with stiffness constant K = 67 kN m^{-1} control the movements in-plane of the other vertices.

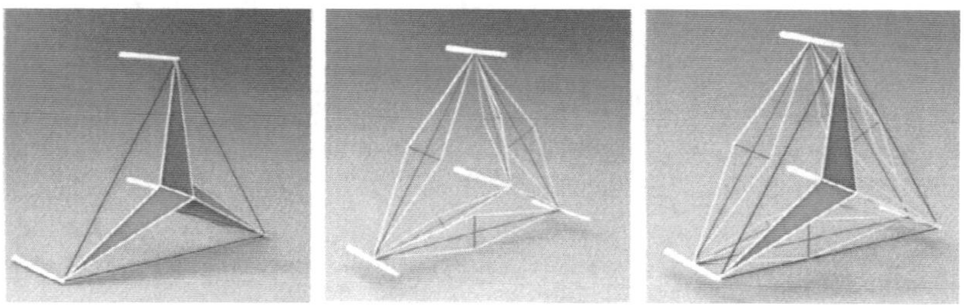

Figure 1. (a) TABS configuration; (b) D-bar elements; (c) construction of the WTABS system.

3 Computational modeling of the WTABS dynamics

In this section, we describe an algorithm for the rigid body dynamics, in matrix form, that includes the cable masses and bar length constraints and handles a rigidity constraint on the deformation of the bars. We will indicate matrices with bold capital letters (i.e. **X**), diagonal matrices with letters with a superimposed hat (i.e. $\hat{x}$), vectors with bold lower-case letters (i.e. **x**), and scalars with italic letters (i.e. x).

For this purpose, we consider a general tensegrity network made of n_n nodes, n_b bars and n_s strings. By introducing the force densities parameter of the kth cable as:

$$\gamma_k = \max\left[0, k_k\left(1-\frac{L_k}{s_k}\right)\right]+\gamma_{ck}, \text{ where } \begin{array}{ll} \gamma_{ck} = c_k\dot{s}_k/s_k & if: \; s_k \geq L_k, \\ \gamma_{ck} = 0, & if: \; s_k < L_k. \end{array} \tag{1}$$

being γ_{ck} the force density due to damping, we can write the equations of motion of a class 1 tensegrity system as follows [10]:

$$\ddot{\mathbf{N}}\mathbf{M} + \mathbf{N}\mathbf{K} = \mathbf{W} \tag{2}$$

where **M** and **K** are respectively the mass matrix and the stiffness matrix defined as

$$\mathbf{M} = \frac{1}{12}\mathbf{C}_\mathbf{B}^T\hat{m}\mathbf{C}_\mathbf{B} + \mathbf{C}_\mathbf{R}^T\hat{m}\mathbf{C}_\mathbf{R}\,, \quad \mathbf{K} = \mathbf{C}_\mathbf{S}^T\hat{\gamma}\mathbf{C}_\mathbf{s} - \mathbf{C}_\mathbf{B}^T\hat{\lambda}\,\mathbf{C}_\mathbf{B} \tag{3}$$

Here $\mathbf{C_B}$, $\mathbf{C_S}$ and $\mathbf{C_R}$ are the connectivity matrices of bars, cables and centers of mass and λ represent the force density in bars per unit length:

$$-\hat{\lambda} = \frac{1}{12}\lfloor\dot{\mathbf{B}}^T\dot{\mathbf{B}}\rfloor\hat{m}\,\hat{l}^{-2} + \frac{1}{2}\lfloor\mathbf{B}^T(\mathbf{W} - \mathbf{S}\hat{\gamma}\mathbf{C}_\mathbf{s})\mathbf{C}_\mathbf{B}^T\rfloor\,\hat{l}^{-2} \tag{4}$$

where $\hat{l}^{-2}$ is a diagonal matrix of terms $l_k^{-2} = \|\mathbf{b}_k\|^{-2}$ (see [8] for further details).

In the field of form finding tensegrity methods, the force density method is used to convert the nonlinear equilibrium systems into linear ones through the transformation between force density (the control variable that appears linearly in the system) and rest length. We apply fourth-order Runge-Kutta method to integrate equation (2) [11]-[13].

4 Numerical applications

The dynamics of the system and the corresponding amount of convertible mechanical energy are investigated under two basic processes, namely motion driven by ram activation and wind fluctuations, due to wind gusts loading the deployed screen by considering the damping factor of 5%, a time step of 0.025 s.

4.1 Simulation of the activation motion of the WTABS module

In this section we assume that the screen is activated by a force acting on a node, that we call as the *actuated node*, impressed by the linear actuator. The loading program, divided in two phases of 40 seconds each, consists in deploying the screen from the initial state to the almost flat configuration, by linearly increasing over time the actuating force, and then maintaining the screen deployed, with the force held constant over time. At t=0 s, the force on the actuated node has value F_{min}=72.690 kN, necessary to sustain the initially pre-stressed configuration. At t=20 s, the external force reaches its maximum value, set to F_{max}=276.221 kN. Notice that, in order to save energy in an actual application, when the screen is folded or deployed at the desired configuration, the actuated node could be locked with a releasable hook and the actuator could be switched off. Whenever the screen configuration must be changed, actuator could be switched on and the hook released. The initial configuration of the system corresponds to the screen almost folded, with initial velocities set to zero. In the chosen initial state, boundary strings are under tensile stress, which is computed as:

$$N_0 = \frac{E_S A_S}{L_0}(L_i - L_0) = 11846.6 \text{ N} \tag{5}$$

where E_s and $A_s = \pi\,(d/2)^2$ are effective Young's modulus and cross-sectional area of the string and L_0 and L_i are rest and initial lengths, respectively given by:

$$L_0 = \frac{\sqrt{2}}{2}L = 3.94042 \text{ m} \qquad \text{and} \qquad L_i = 4.09281 \text{ m} \tag{6}$$

Notice that the rest length is equal to the distance of the boundary nodes of the screen in its fully folded state, which is a purely theoretical and not achievable configuration. This, indeed, ensures that boundary strings in actual applications cannot be stress-free. In Figure 2, the solution is reported with respect to three deformed configurations, namely the initial state, an intermediate level of unfolding, at t = 25 s, and the deployed configuration.

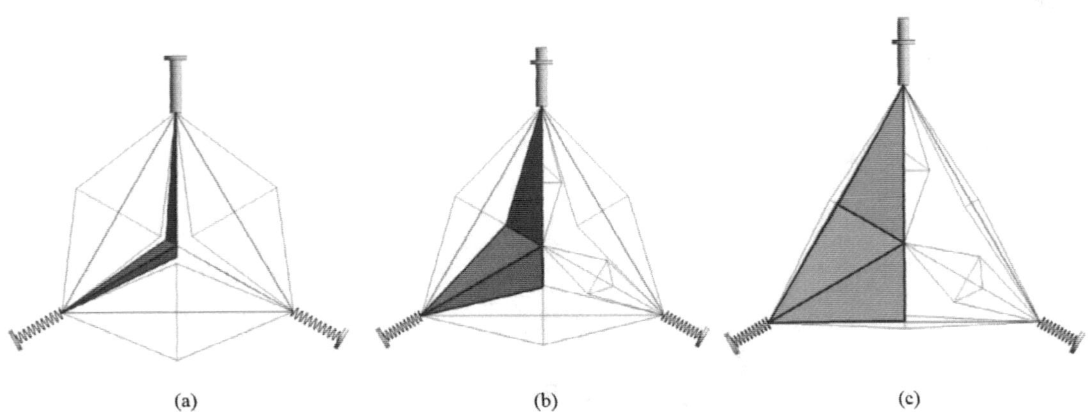

Figure 2. Three configuration of screen and underlying assemblage of D-bars: at t = 0 s (a), t = 25 s (b), and t = 50 s (c). The boundary springs and the actuator ram are represented with amplified scales.

4.2 Modeling of the wind forces

The instantaneous wind vector speed **V**(P,t) can be represented by the sum of the average speed $\mathbf{V}_m$(P) over an interval of 10 min, characterized by long-term variations, and the atmospheric turbulence **V**′(P,t), with high frequency fluctuations [14]:

$$\mathbf{V}_m(P) = \mathbf{i} \cdot v_m(z) \tag{7}$$

$$\mathbf{V}'(P,t) = \mathbf{i} \cdot v'_1(P,t) + \mathbf{j} \cdot v'_2(P,t) + \mathbf{k} \cdot v'_3(P,t) \tag{8}$$

where P is any point at the altitude z on the ground, t is the time, **i**, **j**, **k** are the unit vectors along axes x, y, z; v_m is the average speed along x and v'_1, v'_2, v'_3 are the longitudinal (x), lateral (y) and vertical (z) component of turbulence. We consider only the value of v'_1 (v' from now on) specified by the spectral density [14]:

$$S_L(z,n) = \frac{6.8\, f_L(z,n)}{(1 + 10.2\, f_L(z,n))^{5/3}} \tag{9}$$

where $f_L(z,n)$ is a dimensionless frequency

$$f_L(z,n) = \frac{nL(z)}{v_m(z)} \tag{10}$$

being $L(z)$ the turbolence scale and n the natural frequency.

By selecting a frequency step Δn=0.05 Hz, and a set of 100 frequencies $n_k=(k-1/2)\,\Delta n$, k=1,...,100, we generate a sample of the variable wind speed using the superimposition method with sinusoidal functions, that is [15]:

$$v'(z,t) = \sum_{k=1}^{N} \sqrt{2\, S_L(z,n_k)\Delta n}\, \cos(2\pi n_k t + \varphi_k) \tag{11}$$

with φ_k randomly generated phase angles over the range 0-2π. Finally, the peak value of the wind kinetic pressure, in m/s, is expressed by:

$$q_p(z,t) = \frac{1}{2}\rho\big(v_m(z) + v'(z,t)\big)^2 \tag{12}$$

being ρ=1.25 kg m^{-3} the density of the air.

For the sake of computation, we assume data are compatible with Al Bahar Towers, that is height z^*=120 m in a caostal area exposed to wind, leading to the turbolence scale $L(z^*)$=40.1589 m.

Further, to simulate the wind pressure of 3.5 kPa obtained in wind tunnel tests on full-scale screen prototypes presented in [4] we consider the average velocity $v_m(z^*)$=73.4 m s^{-1}.

The wind force over the generic infill panel is assumed acting along the normal to the panel. Initial configuration coincides with the deployed state after the actuation process. Initial nodal velocities are set to zero. Total time of computation is 250 s.

The time history of the height of the central node of the screen module under wind forces is shown in Figure 3 for a time window of 60 s.

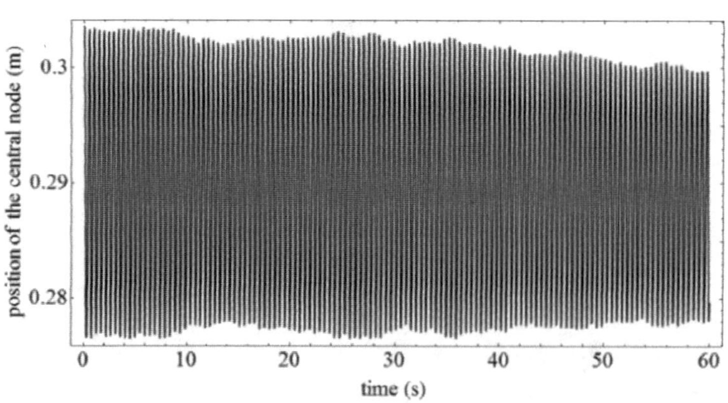

Figure 3. Fluctuation of the height of the central node vs time in the first 60 s of simulation.

IOP Conf. Series: Materials Science and Engineering **999** (2020) 012003 doi:10.1088/1757-899X/999/1/012003

It has been observed in [8] that the overall electric energy from WTABS mounted on a façade like that of Al Bahar Towers during actuation is approximately equal to that produced by 233 PV panels and 87 microelolic rooftop turbines with 1 m^2 surface area. The unit cell of the analyzed sunscreen can also produce 2.3 milliwatt hour in presence of wind forces, and such power can be usefully employed to operate microelectronic devices [16].

5. Concluding remarks

The present work has provided comprehensive computational models in matrix and vector forms for the simulation of the dynamics of dynamic solar façades with a tensegrity-based origami architecture. The special ability of such structures to serve as energy harvesters of new generation energy efficient buildings has been diffusely discussed, by considering their actuation motion and wind induced fluctuations. It has been observed that WTABS has good potential for harvesting the energy. Future studies will be focused on the design of modeling of dynamic facades to develop hybrid structures based on natural ventilation and solar shading elements. Energy harvesting researches will be directed towards the use of various piezoelectric elements, such as ceramics, composites, monocrystals and polymers, for the electrical conversion of the energy. The experimental validation of the analyzed sunscreens will be conducted on reduced scale physical models through structural identification procedures.

Acknowledgements
R.M., N.S., and F.F. acknowledge financial support from MIUR under the PRIN 2017 National Grant 'Multiscale Innovative Materials and Structures' (grant number 2017J4EAYB).

References
[1] European Commission, HORIZON 2020 Work Programme 2014-2015, PART 5. II, p 98.
[2] Babilio E, Miranda R and Fraternali F 2019 *Frontiers in Materials* **6**, Article no. 7, pp 48-59.
[3] Armstrong A, Buffoni G, Eames D, James R, Lang L, Lyle J and Xuereb K 2013 *Arup J* **2**, pp 90-95.
[4] Karanouh A and Kerber E 2015 *Journal of Façade Design and Engineering* **3(2)**, pp 185–221.
[5] Balduzzi F, Bianchini A and Ferrari L 2012 *Renewable Energy* **5**, pp.163–174.
[6] Gagliano A, Nocera F, Patania F and Capizzi A. 2013 *Int J Energy Environ Eng* **4** pp 43.
[7] Kosasih B and Tondelli A, 2012 *Procedia Engineering*, *49*, pp 92-98.
[8] Miranda R, Babilio E, Singh N, Santos F and Fraternali F 2020 *Mech Res Commun* **105**, Article no. 103503.
[9] Power Jacks, Rolaram, Electric Linear Actuators. Available at: https:// www.powerjacks.com /perch/resources/brochure/pjlab-rolaram-en-01-1b.pdf.
[10] Skelton R E: 2005 *IUTAM Symposium on Vibration Control of Nonlinear Mechanisms and Structures*, Springer Netherlands pp. 309–318.
[11] Fabbrocino F and Carpentieri G 2017 *Composite Structures* **173**, pp 9–16.
[12] Nagase K and Skelton R E 2014 *Mech Res Commun* **59**, pp 14–25.
[13] Xu X, Li S and Luo Y. 2017 *Mech Res Commun* **84**, pp 1-7.
[14] EN 1991-1-4 (2005): Eurocode 1: Actions on structures, Parts 1-4: General actions. Wind actions.
[15] Santos F, Gonçalves P, Cismaşiu C and Gamboa-Marrufo M 2014 *Proceedings of the Institution of Civil Engineers-Structures and Buildings* **167**(12), pp 743-52.
[16] De Pasquale G 2013 *In Handbook of MEMS for wireless and mobile applications* (Woodhead Publishing) p 345-400.

IOP Conf. Series: Materials Science and Engineering **999** (2020) 012004 doi:10.1088/1757-899X/999/1/012004

Formation of Discontinuities in Rectangular Plates as a Result of Residual Stress Relief

I V Menshova[1,2,*], A P Kerzhaev[1], G Yu[3,4] and X Zeng[3,4]

[1] Institute of Earthquake Prediction Theory and Mathematical Geophysics, Russian Academy of Sciences, Moscow 117997, Russia
[2] Bauman Moscow State Technical University, Moscow 105005, Russia
[3] School of Civil Engineering, Qingdao University of Technology, Qingdao 266033, China
[4] Cooperative Innovation Center of Engineering Construction and Safety in Shandong Blue Economic Zone Qingdao University of Technology, Qingdao 266033, China

[*]E-mail: menshovairina@yandex.ru

Abstract. The paper deals with the problem of relieving residual stresses in an elastic domain of rectangular shape with free sides as a result of the formation of a discontinuity of particular shape. First, we construct the solution to the problem of residual stresses in an infinite strip with free sides and with a central transverse cut on which a discontinuity of displacements is known. Then, the solution for a rectangle is added to this solution, with the help of which the boundary conditions at the ends are satisfied. The formulas for the residual stresses and for the corresponding displacements are represented in the form of series in Papkovich–Fadle eigenfunctions. The expansion coefficients (Lagrange coefficients) have the form of simple Fourier integrals.

1. Introduction

Accounting for residual stresses in engineering is one of the most difficult scientific problems. We note two books that, in our opinion, deserve the greatest attention from the point of view of practice. The first book [1] is the result of many years of research on various manifestations of residual stresses in the construction and operation of underground structures. In the second book [2], various physical models of residual stresses are constructed.

The determination of residual stresses, particularly in rock samples, is a very difficult problem. Simple initial estimates of the magnitude and distribution laws of residual stresses can be made on the basis of the analysis of the shape and location of cracks formed during residual stress relief, as well as the solution of the corresponding model problems of the type considered in the paper.

Based on the mathematical theory of residual stresses [3–5], in this paper, three types of discontinuities in a rectangular domain were considered, from which preliminary conclusions about residual stresses can be made. Then, these estimates can be refined by solving the corresponding boundary value problems of the theory of elasticity.

The obtained formulas, strictly speaking, describe residual stresses in a rectangular plate and the displacements that occur as a result of their relief with the formation of discontinuities. Therefore, the presentation is conducted in terms of residual stresses. In order to pass to the solutions in the classical formulation (a rectangle with a crack), it is necessary to change the signs in the formulas for displacements or stresses to the opposite ones in a certain way.

The problems of rectangular plates weakened by cracks have been the subject of numerous studies. In the paper [6], the stress intensity factors were investigated for an infinite elastic strip containing a central transverse crack. Reut, Vaysfeld and Zhuravlova [7, 8] investigated the stress state of a semi-strip with a transverse and longitudinal crack. The solution to the problem is reduced to solving a system of three singular integral equations. Their articles also provide an extensive review of publications on this topic. In the papers [9, 10], numerical methods are used to investigate the problems of a central crack in a rectangular plate subjected to external tension. Zhou and Yang [11] used the near crack line analysis method to investigate a central crack loaded by two pairs of point shear forces in a finite plate and obtained the analytical solution.

2. Solution in a strip

Let us consider the infinite free strip $\{\Pi : |x| < \infty, |y| \le 1\}$ in which there is a field of residual stresses symmetric with respect to the x and y axes. We make the transverse cut $\{\gamma : x = 0, |y| \le \alpha < 1\}$ and assume that a discontinuity of longitudinal displacements is given on this cut:

$$U^{+}(0,y) - U^{-}(0,y) = 2u(y), \tag{1}$$

where $U^{\pm}(0,y)$ denotes the displacements equal to $u(y)$ on the right and left of the cut, respectively.

We will seek the solution to the problem in the form of series $(x \ge 0,\ \mathrm{Re}\,\lambda_k < 0)$

$$\begin{aligned}
U(x,y) &= \sum_{k=1}^{\infty} a_k \xi(\lambda_k, y) e^{\lambda_k x} + \overline{a_k} \xi(\overline{\lambda_k}, y) e^{\overline{\lambda_k} x},\\
V(x,y) &= \sum_{k=1}^{\infty} a_k \chi(\lambda_k, y) e^{\lambda_k x} + \overline{a_k} \chi(\overline{\lambda_k}, y) e^{\overline{\lambda_k} x},\\
\sigma_x(x,y) &= \sum_{k=1}^{\infty} a_k s_x(\lambda_k, y) e^{\lambda_k x} + \overline{a_k} s_x(\overline{\lambda_k}, y) e^{\overline{\lambda_k} x},\\
\sigma_y(x,y) &= \sum_{k=1}^{\infty} a_k s_y(\lambda_k, y) e^{\lambda_k x} + \overline{a_k} s_y(\overline{\lambda_k}, y) e^{\overline{\lambda_k} x},\\
\tau_{xy}(x,y) &= \sum_{k=1}^{\infty} a_k t_{xy}(\lambda_k, y) e^{\lambda_k x} + \overline{a_k} t_{xy}(\overline{\lambda_k}, y) e^{\overline{\lambda_k} x}
\end{aligned} \tag{2}$$

in the Papkovich–Fadle eigenfunctions

$$\begin{aligned}
\xi(\lambda_k, y) &= \left(\frac{1-\nu}{2} \sin\lambda_k - \frac{1+\nu}{2} \lambda_k \cos\lambda_k \right) \cos\lambda_k y - \frac{1+\nu}{2} \lambda_k y \sin\lambda_k \sin\lambda_k y,\\
\chi(\lambda_k, y) &= \left(\frac{1+\nu}{2} \lambda_k \cos\lambda_k + \sin\lambda_k \right) \sin\lambda_k y - \frac{1+\nu}{2} \lambda_k y \sin\lambda_k \cos\lambda_k y,\\
s_x(\lambda_k, y) &= (1+\nu)\lambda_k \left[(\sin\lambda_k - \lambda_k \cos\lambda_k) \cos\lambda_k y - \lambda_k y \sin\lambda_k \sin\lambda_k y \right],\\
s_y(\lambda_k, y) &= (1+\nu)\lambda_k \left[(\sin\lambda_k + \lambda_k \cos\lambda_k) \cos\lambda_k y + \lambda_k y \sin\lambda_k \sin\lambda_k y \right],\\
t_{xy}(\lambda_k, y) &= (1+\nu)\lambda_k^2 (\cos\lambda_k \sin\lambda_k y - y \sin\lambda_k \cos\lambda_k y).
\end{aligned} \tag{3}$$

In formulas (2) and (3), the following notation is introduced: $U(x,y)$ and $V(x,y)$ are the longitudinal and transverse displacements, respectively, multiplied by the shear modulus G, ν is Poisson's ratio, a_k are the unknown expansion coefficients, the eigenvalues λ_k are the zeros of the entire function of exponential type $L(\lambda) = \lambda + \sin\lambda\cos\lambda$.

The expansion coefficients a_k in formulas (2) are determined from the conditions on the crack as a solution to the problem of conjugation of two functions analytical in the right and left semi-strips. The meaning of their introduction is to extract minimal systems of functions, to which it is possible to construct biorthogonal systems of functions, from the right and left complete systems of the Papkovich–Fadle eigenfunctions involved in the boundary conditions at the junction of the semi-strips, and then, with the help of them, to find the required coefficients of expansions.

To formulate the boundary value problem for an infinite strip with the discontinuity of longitudinal displacements (1), following the papers [3, 4, 12], we introduce two functions:

$$\begin{aligned} \Phi(x,y) &= -\tau_{xy}(x,y) + i[(1+\nu)\frac{\partial V(x,y)}{\partial y} - \frac{1-\nu}{2}\sigma_x(x,y)], \\ F(x,y) &= -(1+\nu)\frac{\partial U(x,y)}{\partial y} - \frac{1-\nu}{2}\tau_{xy}(x,y) - i\sigma_x(x,y). \end{aligned} \tag{4}$$

Substituting here (2), we obtain

$$\Phi(x,y) = \sum_{k=1}^{\infty} 2\operatorname{Re}\left[a_k \Phi(\lambda_k, y) e^{\lambda_k x}\right],\ F(x,y) = \sum_{k=1}^{\infty} 2\operatorname{Re}\left[a_k F(\lambda_k, y) e^{\lambda_k x}\right], \tag{5}$$

where

$$\begin{aligned} \Phi(\lambda_k, y) &= (1+\nu)\lambda_k^2 (i\cos\lambda_k + y\sin\lambda_k) e^{i\lambda_k y}, \\ F(\lambda_k, y) &= (1+\nu)\lambda_k [i(\lambda_k\cos\lambda_k - \sin\lambda_k) + \lambda_k y \sin\lambda_k] e^{i\lambda_k y}. \end{aligned} \tag{6}$$

Denote

$$\omega_1 = \lambda_1,\ \omega_2 = -\lambda_1,\ \omega_3 = \lambda_2,\ \omega_4 = -\lambda_2,\ \dots,\ a_k^+ = A_k,\ a_k^- = -A_k, \tag{7}$$

where a_k^+, a_k^- are the unknown expansion coefficients for the right and left semi-strips, respectively.

Then, from the conjugation conditions for $x=0$, we obtain two equations:

$$\sum_{k=1}^{\infty} 2\operatorname{Re}[A_k \Phi(\omega_k, y)] = 0,\ \sum_{k=1}^{\infty} 2\operatorname{Re}[A_k F(\omega_k, y)] = 2\frac{du(y)}{dy}. \tag{8}$$

The solution to the system of equations (8) is constructed in the same way as in [3], namely with the help of the functions $\Phi_k(y)$ and $F_k(y)$ biorthogonal to the functions $\Phi(\lambda_k, y)$ and $F(\lambda_k, y)$. The biorthogonal functions are determined from the equations [3]

$$\int_{-\infty}^{\infty} \Phi(\lambda, y)\Phi_k(y)dy = \frac{\lambda L(\lambda)}{\lambda - \lambda_k},\ \int_{-\infty}^{\infty} F(\lambda, y)F_k(y)dy = \frac{L(\lambda)}{\lambda - \lambda_k}. \tag{9}$$

Using the biorthogonality relations following from equations (9), from (8) we obtain the system of two algebraic equations

$$\begin{cases} A_k \omega_k M_k + \overline{A_k}\,\overline{\omega_k}\,\overline{M_k} = 0; \\ A_k M_k + \overline{A_k}\,\overline{M_k} = f_k + \overline{f_k}. \end{cases} \tag{10}$$

Here

$$f_k = \frac{u_k}{\lambda_k},\ u_k = \int_{-\alpha}^{\alpha} u(y) u_k(y) dy,\ u_k(y) = \frac{\lambda_k \cos\lambda_k y}{(1+\nu)\sin\lambda_k}. \tag{11}$$

The numbers u_k are the Lagrange coefficients of the expanded function $u(y)$, the functions $u_k(y)$ are the finite parts of the functions biorthogonal to the Papkovich–Fadle eigenfunctions $\xi(\lambda_k, y)$, $M_k = \cos^2\lambda_k$ are normalizing factors.

Solving the system of equations (10) and taking into account the notation (7), we find

$$a_k^+ = -\frac{\overline{\lambda_k}(f_k + \overline{f_k})}{(\lambda_k - \overline{\lambda_k})M_k}. \tag{12}$$

We substitute (12) into (2) and separate the zero-series, following [3, 4], for example. As a result, we obtain formulas that describe the solution to the problem under consideration:

$$U^s(x,y) = -\operatorname{sgn}(x)\sum_{k=1}^{\infty} 2\operatorname{Re}\left\{u_k \frac{\xi(\lambda_k, y)}{\lambda_k M_k}\frac{\operatorname{Im}(\overline{\lambda_k} e^{\lambda_k |x|})}{\operatorname{Im}\lambda_k}\right\},$$

$$V^s(x,y) = -\sum_{k=1}^{\infty} 2\operatorname{Re}\left\{\frac{u_k}{\lambda_k}\frac{\chi(\lambda_k, y)}{M_k}\frac{\operatorname{Im}(\overline{\lambda_k} e^{\lambda_k |x|})}{\operatorname{Im}\lambda_k}\right\},$$

$$\sigma_x^s(x,y) = -\sum_{k=1}^{\infty} 2\,\mathrm{Re}\left\{ \frac{u_k}{\lambda_k} \frac{s_x(\lambda_k, y)}{M_k} \frac{\mathrm{Im}(\overline{\lambda_k} e^{\lambda_k |x|})}{\mathrm{Im}\,\lambda_k} \right\}, \tag{13}$$

$$\sigma_y^s(x,y) = -\sum_{k=1}^{\infty} 2\,\mathrm{Re}\left\{ \frac{u_k}{\lambda_k} \frac{s_y(\lambda_k, y)\lambda_k \overline{\lambda_k}}{M_k \lambda_k^2} \frac{\mathrm{Im}(\lambda_k e^{\lambda_k |x|})}{\mathrm{Im}\,\lambda_k} \right\},$$

$$\tau_{xy}^s(x,y) = -\mathrm{sgn}(x) \sum_{k=1}^{\infty} 2\,\mathrm{Re}\left\{ u_k \overline{\lambda_k} \frac{t_{xy}(\lambda_k, y)}{\lambda_k M_k} \frac{\mathrm{Im}(e^{\lambda_k |x|})}{\mathrm{Im}\,\lambda_k} \right\}.$$

Formulas (13) describe the residual stresses in the infinite strip Π and the displacements that occur during their relief with the formation of the discontinuity (1).

3. Solution in a rectangle

The solution for the free rectangle $\{\mathrm{P} : |x| \le d, |y| \le 1\}$ with the discontinuity (1) is obtained simply. To the solution (13) for the strip with the discontinuity, we need to add the solution for the rectangle that relieves the normal and tangential stresses

$$\sigma_x^s(\pm d, y) = -\sum_{k=1}^{\infty} 2\,\mathrm{Re}\left\{ \frac{u_k}{\lambda_k} \frac{s_x(\lambda_k, y)}{M_k} \frac{\mathrm{Im}(\overline{\lambda_k} e^{\lambda_k |d|})}{\mathrm{Im}\,\lambda_k} \right\},$$

$$\tau_{xy}^s(\pm d, y) = -\mathrm{sgn}(\pm d) \sum_{k=1}^{\infty} 2\,\mathrm{Re}\left\{ u_k \overline{\lambda_k} \frac{t_{xy}(\lambda_k, y)}{\lambda_k M_k} \frac{\mathrm{Im}(e^{\lambda_k |d|})}{\mathrm{Im}\,\lambda_k} \right\} \tag{14}$$

in the cross sections $x = \pm d$ of the strip.

The final formulas for the displacements and stresses in the rectangle when normal and tangential stresses are given at its ends in the case of its even-symmetric deformation with respect to the central axes are as follows:

(a) if only normal stresses are given at the ends of the rectangle, then

$$U(x,y) = \sum_{k=1}^{\infty} 2\,\mathrm{Re}\left(\sigma_k \overline{\lambda_k} \frac{\xi(\lambda_k, y)}{\lambda_k M_k} \right) \frac{\mathrm{Im}(\lambda_k \sinh \overline{\lambda_k} d \sinh \lambda_k x)}{\mathrm{Im}(\overline{\lambda_k} \sinh \overline{\lambda_k} d \cosh \lambda_k d)},$$

$$V(x,y) = \sum_{k=1}^{\infty} 2\,\mathrm{Re}\left(\sigma_k \frac{\chi(\lambda_k, y)}{M_k} \right) \frac{\mathrm{Im}(\overline{\lambda_k} \sinh \overline{\lambda_k} d \cosh \lambda_k x)}{\mathrm{Im}(\overline{\lambda_k} \sinh \overline{\lambda_k} d \cosh \lambda_k d)},$$

$$\sigma_x(x,y) = \sum_{k=1}^{\infty} 2\,\mathrm{Re}\left(\sigma_k \frac{s_x(\lambda_k, y)}{M_k} \right) \frac{\mathrm{Im}(\overline{\lambda_k} \sinh \overline{\lambda_k} d \cosh \lambda_k x)}{\mathrm{Im}(\overline{\lambda_k} \sinh \overline{\lambda_k} d \cosh \lambda_k d)}, \tag{15}$$

$$\sigma_y(x,y) = \sum_{k=1}^{\infty} 2\,\mathrm{Re}\left(\sigma_k \frac{s_y(\lambda_k, y)}{\lambda_k^2 M_k} \right) \frac{\mathrm{Im}(\lambda_k \sinh \overline{\lambda_k} d \cosh \lambda_k x)}{\mathrm{Im}(\overline{\lambda_k} \sinh \overline{\lambda_k} d \cosh \lambda_k d)} \lambda_k \overline{\lambda_k},$$

$$\tau_{xy}(x,y) = \sum_{k=1}^{\infty} 2\,\mathrm{Re}\left(\sigma_k \frac{t_{xy}(\lambda_k, y)}{\lambda_k M_k} \right) \frac{\mathrm{Im}(\sinh \overline{\lambda_k} d \sinh \lambda_k x)}{\mathrm{Im}(\overline{\lambda_k} \sinh \overline{\lambda_k} d \cosh \lambda_k d)} \lambda_k \overline{\lambda_k};$$

(b) if only tangential stresses are given at the ends of the rectangle, then

$$U(x,y) = -\sum_{k=1}^{\infty} 2\,\mathrm{Re}\left(\tau_k \frac{\xi(\lambda_k, y)}{\lambda_k M_k} \right) \frac{\mathrm{Im}(\lambda_k \cosh \overline{\lambda_k} d \sinh \lambda_k x)}{\mathrm{Im}(\lambda_k \sinh \lambda_k d \cosh \overline{\lambda_k} d)},$$

$$V(x,y) = -\sum_{k=1}^{\infty} 2\,\mathrm{Re}\left(\frac{\tau_k}{\lambda_k} \frac{\chi(\lambda_k, y)}{M_k} \right) \frac{\mathrm{Im}(\lambda_k \cosh \overline{\lambda_k} d \cosh \lambda_k x)}{\mathrm{Im}(\lambda_k \sinh \lambda_k d \cosh \overline{\lambda_k} d)},$$

$$\sigma_x(x,y) = -\sum_{k=1}^{\infty} 2\,\mathrm{Re}\left(\tau_k \frac{s_x(\lambda_k, y)}{M_k} \right) \frac{\mathrm{Im}(\cosh \overline{\lambda_k} d \cosh \lambda_k x)}{\mathrm{Im}(\lambda_k \sinh \lambda_k d \cosh \overline{\lambda_k} d)}, \tag{16}$$

$$\sigma_y(x,y) = -\sum_{k=1}^{\infty} 2\,\mathrm{Re}\left(\tau_k \frac{s_y(\lambda_k, y)}{\lambda_k^2 M_k} \right) \frac{\mathrm{Im}(\lambda_k^2 \cosh \overline{\lambda_k} d \cosh \lambda_k x)}{\mathrm{Im}(\lambda_k \sinh \lambda_k d \cosh \overline{\lambda_k} d)},$$

$$\tau_{xy}(x,y) = -\sum_{k=1}^{\infty} 2\,\mathrm{Re}\left(\tau_k \frac{t_{xy}(\lambda_k, y)}{\lambda_k M_k}\right) \frac{\mathrm{Im}(\lambda_k \cosh \overline{\lambda_k} d \sinh \lambda_k x)}{\mathrm{Im}(\lambda_k \sinh \lambda_k d \cosh \overline{\lambda_k} d)}.$$

The numbers σ_k and τ_k should be taken equal to

$$\sigma_k = \frac{u_k}{\lambda_k} \frac{\mathrm{Im}(\overline{\lambda_k} e^{\lambda_k |d|})}{\mathrm{Im}\,\lambda_k},\ \tau_k = -u_k \overline{\lambda_k} \frac{\mathrm{Im}(e^{\lambda_k |d|})}{\mathrm{Im}\,\lambda_k}. \tag{17}$$

The complete solution to the problem for the rectangle with the discontinuity of longitudinal displacements (1) is obtained by adding the solutions (13), (15) and (16).

Let us consider three types of discontinuities of longitudinal displacements:

$$u_1(y) = \begin{cases} \sqrt{\alpha^2 - y^2}, & |y| < \alpha < 1; \\ 0, & \alpha < |y| \le 1, \end{cases} \quad u_2(y) = \begin{cases} 2(\alpha^2 - y^2), & |y| < \alpha < 1; \\ 0, & \alpha < |y| \le 1, \end{cases} \quad u_3(y) = \begin{cases} 8(\alpha^2 - y^2)^2, & |y| < \alpha < 1; \\ 0, & \alpha < |y| \le 1. \end{cases} \tag{18}$$

Accept $\alpha = 0.5$. Using formulas (11) and (17), we find the Lagrange coefficients u_k, σ_k and τ_k for each of the functions (18), and then substitute them into (13), (15) and (16). Figures 1 and 2 show the comparison of the three solutions depending on the shape of the discontinuity. The solid lines correspond to the solution for the discontinuity $u_1(y)$, the dashed lines for $u_2(y)$, and the dotted lines for $u_3(y)$.

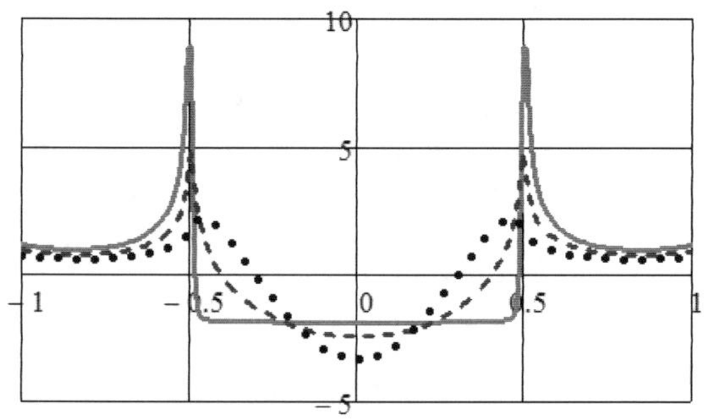

Figure 1. Distribution of the normal stresses $\sigma_x(0.01, y)$ depending on the shape of the discontinuity.

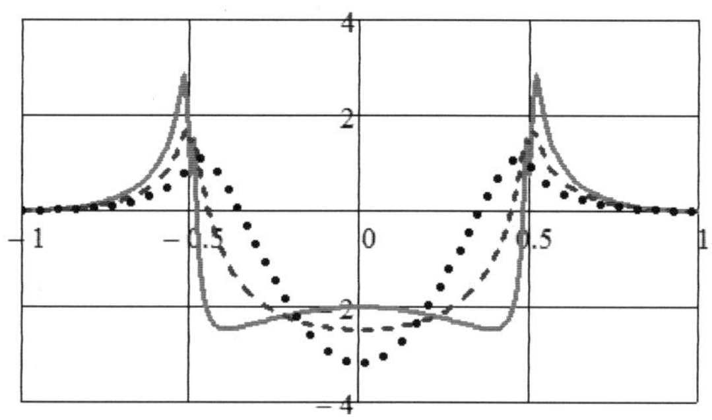

Figure 2. Distribution of the normal stresses $\sigma_y(0.01, y)$ depending on the shape of the discontinuity.

4. Conclusions

The paper proposes a new method for solving the problem of relieving residual stresses in an elastic rectangular domain with a transverse crack on which a displacement discontinuity is given. It is based on the theory of expansions in Papkovich–Fadle eigenfunctions. Three types of discontinuities in the rectangular domain were considered. The comparison of the three solutions was given depending on the shape of the discontinuity. The solutions obtained make it possible to reconstruct, from the shape of the discontinuity, the fields of residual stresses that led to their formation.

The problem is somewhat more complicated when stresses are given on the crack. Using the same scheme, one can obtain similar solutions for a rectangle with discontinuities of displacements or with stresses given at the discontinuity with other boundary conditions on its sides, particularly with stiffeners. Using the simple superposition method, one can construct solutions for a set of discontinuities if their shapes and locations are known. In the future, the authors intend to construct similar exact solutions for finite bodies of canonical shape in oblique and polar coordinate systems, in particular, for a rectangle with an oblique crack. This problem has an important application in the problem of increasing the recovery of oil-bearing layers.

Acknowledgments
The research was funded by RFBR and NSFC according to the research project No. 20-51-53021, by the Russian Science Foundation under grant No. 19-71-00094, and by the National Natural Science Foundation of China (No. 51674150) and NSFC-RFBR (No. 5191101589).

References

[1] Tazhibaev K T 2016 *Stresses, Processes of Deformation and Dynamic Destruction of Rocks* In two volumes (Bishkek: Altyn Print) (in Russian)
[2] Moroz A I 2004 *Self-Stressed State of Rocks* (Moscow: MGGU) (in Russian)
[3] Kovalenko M D, Menshova I V and Shulyakovskaya T D 2013 Expansions in Fadle–Papkovich functions: examples of solutions in a half-strip *Mech. Solids* **48**(5) 584–602
[4] Kovalenko M D, Menshova I V and Kerzhaev A P 2018 On the exact solutions of the biharmonic problem of the theory of elasticity in a half-strip *Z. Angew. Math. Phys.* **69** 121
[5] Kovalenko M D, Abrukov D A, Menshova I V, Kerzhaev A P and Yu G 2019 Exact solutions of boundary value problems in the theory of plate bending in a half-strip: basics of the theory *Z. Angew. Math. Phys.* **70** 98
[6] Goldstein R V, Ryskov I N and Salganik R L 1970 Central transverse crack in an infinite strip *Int. J. Fract.* **6**(1) 104–5
[7] Reut V, Vaysfeld N and Zhuravlova Z 2019 Investigation of the stress state of the elastic semi-strip with a transverse crack *Theor. Appl. Fract. Mech.* **100** 105–9
[8] Vaysfeld N and Zhuravlova Z 2020 The investigation of semi-strip's stress state with a longitudinal crack *Z. Angew. Math. Mech.* **100**(3) e201900289
[9] Mohsin N R 2015 Static and dynamic analysis of center cracked finite plate subjected to uniform tensile stress using finite element method *Int. J. Mech. Eng. Technol.* **6**(1) 56–70
[10] Li X and You X 2006 Boundary collocation method for a cracked rectangular plate with double external tension *Appl. Anal.* **85**(9) 1103–11
[11] Zhou X P and Yang H Q 2009 Elastoplastic solution for an eccentric crack loaded by two pairs of point tensile forces *Theor. Appl. Fract. Mech.* **51**(1) 62–72
[12] Kovalenko M D, Menshova I V, Kerzhaev A P and Yu G 2018 Mixed boundary value problems in the theory of elasticity in an infinite strip *Acta Mech.* **229**(11) 4339–56

IOP Conf. Series: Materials Science and Engineering **999** (2020) 012005 doi:10.1088/1757-899X/999/1/012005

Lubrication performance analysis of the connecting rod bearing based on AVL EXCITE

Ning Zhong[1], Bo Wu[1], Le Zhang[1], Qi Xin[1], Liyue Tu[1], Hulin Li[2,*]

[1]Chongqing Yuejin Machinery Co., Ltd., No.1 Huagong RD, Yongchuan District, Chongqing 402169, PR China
[2] School of Design, Shanghai Jiao Tong University, 800 Dongchuan RD, Minhang District, Shanghai 200240, PR China
[*]Corresponding author: E-mail address: lhl666@sjtu.edu.cn (Hulin Li)

Abstract. The lubrication state of connecting rod bearing has influence on the reliability and lifetime. It is necessary to analyze the lubrication state of connecting rod bearing to evaluate and design the lubrication performance. The connecting rod bearing hydrodynamic lubrication simulation model was set up for 9XX diesel engine and journal bearing fatigue test-rig at Glacier Vandervall. The effect of maximum oil film pressure, minimum oil film thickness and maximum asperity contact pressure on the lubrication performance were studied. Then the lubrication states of connecting rod bearings under different working conditions were discussed based on two test machines. The results are of great significance for analyzing the lubrication performance of connecting rod bearings and the reliability of bearings.

Keywords: Connecting rod bearings; AVL EXCITE; lubrication state; EHD; Lubrication performance

1. Introduction

As one of the most critical elements in diesel engines, journal bearings are applied to support load and transform the reciprocating motion of the pistons into the rotary motion of the crankshaft [1]. According to the survey, bearing failure incidents have accounted for 24.4% ofengine failure incidents and cost [2]. If the connecting rod bearing runs for a long time in a poor lubrication state, the asperity contact will occur between the bearing and the journal, which causes to severe bearing failure [3]. Therefore, it is necessary to study the lubrication performance of the connecting rod bearing, this can provide guidence for improving the reliability and service life of the bearing in actual industry.

The researches of the bearing lubrication performance analysis usually simplify the geometry model and obtain the lubrication states using the Reynolds equation derived from Navier-Stokes equations and continuity equations [4-5]. However, there are still some limitations as follows: firstly, the lubricant viscosity is assumed to the constant neglecting the influence of viscosity changing on the lubrication performance [6-7]. Secondly, the elastic deformation of the bearing and the journal is not considered in some articles [8]. Furthermore, some researchers ignore the effect of surface topography parameters on lubrication characteristics [9-10]. Therefore, in this paper, considering the factors of the lubricant viscosity, the elastic deformation, and the surface topography, the lubrication performance of the connecting rod bearing in 9XX diesel engine and journal bearing fatigue test-rig at Glacier Vandervall is discussed, the lubrication states of connecting rod bearings under different working conditions in two devices are compared to provide technical support for bearing lubrication performance analysis and the design of the working conditions.

IOP Conf. Series: Materials Science and Engineering **999** (2020) 012005 doi:10.1088/1757-899X/999/1/012005

2. Theory

2.1. Averaged Reynolds equation

The full oil film interaction is solved by the averaged Reynolds equation, which considers the influence of roughness amplitudes. The specific formula is as follows [11]:

$$\frac{\partial}{\partial x}\left(-\frac{h_T^3}{12\eta}\frac{\partial P}{\partial x}+\frac{u_1+u_2}{2}h_T\right)+\frac{\partial}{\partial y}\left(-\frac{h_T^3}{12\eta}\frac{\partial P}{\partial y}\right)+\frac{\partial h_T}{\partial t}=0 \quad (1)$$

where h_T denotes the real clearance considering roughness heights of journal and bearing. P is the hydrodynamic pressure and x, y represents the circumferential and axial direction respectively. H indicates the oil viscosity.

2.2. Lubricating oil properties

The Vogel equation [12] is applied to describe the relationship between the oil viscosity and oil temperature in the simulation model:

$$\eta(T)=Ae^{\frac{B}{T+C}} \quad (2)$$

where A=0.198 mPa·s, B=736.69 °C, C=88.9 °C for oil Shell Rimula R3 Multi 10W-30 in this paper. T is the oil temperature.

2.3. Asperity contact interaction

Asperity contact interaction exists when asperity summits of two nominal flats rub each other. The Greenwood and Tripp model is used to evaluate the behaviour of asperity contact. The asperity contact pressure is calculated as:

$$P_a=KE^*F_{5/2}(H_s) \quad (3)$$

where the elastic factor K is set to 0.003. $F_{5/2}(H_s)$ stands for the form function and E^* is the composite elastic modulus, as listed in Eq. (4) and (5):

$$F_{5/2}(H_s)=\begin{cases}4.4086\times10^{-5}(4-H_s)^{6.804} & H_s<4\\ 0 & H_s\geq 4\end{cases} \quad (4)$$

$$E^*=\frac{1}{\left(\frac{1-\nu_1^2}{E_1}+\frac{1-\nu_2^2}{E_2}\right)} \quad (5)$$

where H_s is the non-dimensional summit clearance, ν_1 and ν_2 are Poisson ratio of bearing material and shaft material respectively, E_1 and E_2 are Young modulus of bearing material and shaft material respectively.

2.4. Lubrication conditions

Generally, the film thickness ratio is designed as a criterion for the determination of lubrication condition. It is calculated as follows:

$$\lambda=\frac{h_{\min}}{\sqrt{\sigma_1^2+\sigma_2^2}} \quad (6)$$

where $h_{\min}$ represents the minimum oil film thickness. σ_1 and σ_2 are surface roughness of bearing and journal respectively. According to the range of the film thickness ratio, journal bearing lubrication regimes are divided into three kinds: boundary lubrication ($\lambda \leq 1$), mixed lubrication ($1 < \lambda < 3$), hydrodynamic lubrication ($\lambda \geq 3$).

IOP Conf. Series: Materials Science and Engineering **999** (2020) 012005 doi:10.1088/1757-899X/999/1/012005

3. Simualtion

The connecting rod big-end bearings of 9XX diesel engine and journal bearing fatigue test-rig at Glacier Vandervall are modelled by AVL EXCITE to study the lubrication performance. The modeling process is as follows: firstly, the geometic model of the connecting rod assembly is established in Solidworks, which includes the connecting rod, the connecting rod small-end bearing, and the connecting rod big-end bearing. Then the geometic model is meshed in hypemesh software. The tetrahedral mesh is used for the connecting rod model and the hexahedral mesh is applied for the small-end and big-end bearing models. Based on the mesh model, the finite element model is obtained by Abaqus solver. Finally, in AVL EXCITE, the connecting rod dynamic model is built and the information of the finite element model is input. The simulation models of two devices are shown in Figure 1. By setting the simulation parameters and adding the hydraulic cylinder pressure, the lubrication characteristics can be obtained. In this paper, the lubrication performance of 9XX diesel engine under two cylinder pressure conditons and journal bearing fatigue test-rig under a maximum cylinder pressure of 40Mpa is calculated. Table 1 lists the simulation parameters. The plots of the hydraulic cylinder pressure of two devices are shown in Figure 2.

Table 1. Simulation parameters.

Parameter	9XX diesel engine	Journal bearing fatigue test-rig at Glacier Vandervall
Hydraulic cylinder type	four-stroke	two-stroke
The number of hydraulic cylinders	20	1
Hydraulic cylinder diameter (mm)	230	130
Stroke (mm)	230	0.76
Shaft rotational speed (rpm)	1405/1455	3000
Shaft diameter (mm)	158	52.7
Bearing width (mm)	62	29.62
Lubricating oil supply pressure (bar)	8	6
Inlet oil temperature (°C)	90	70
Shaft surface roughness (μm)	0.63	0.2
Bearing surface roughness (μm)	0.2	0.45

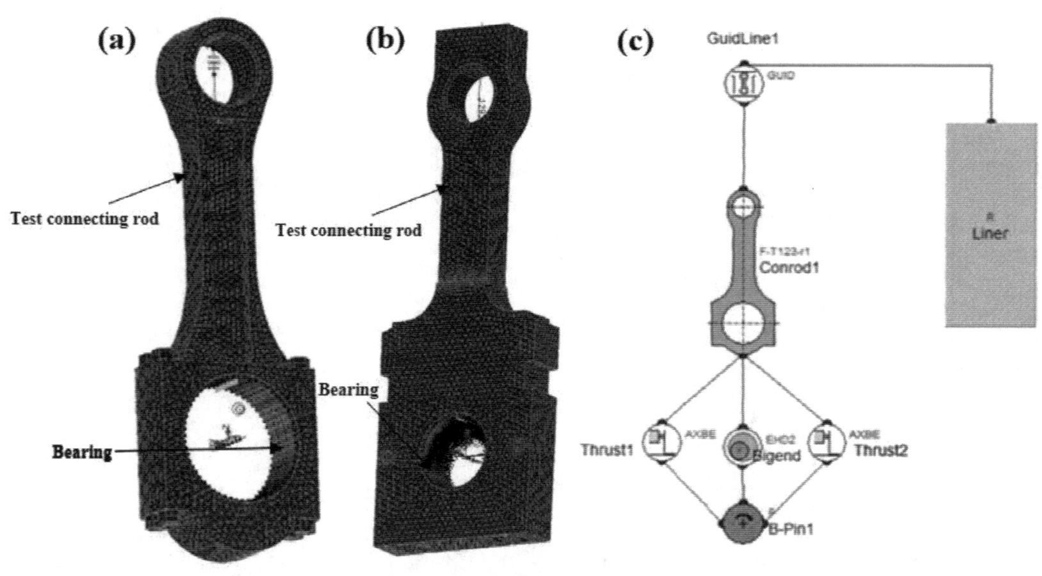

Figure 1. The simulation models: (a): the finite element model of the connecting rod assembly of 9XX diesel engine; (b): the finite element model of the connecting rod assembly of journal bearing fatigue test-rig at Glacier Vandervall; (c): the dynamic model.

4. Results and discussion

4.1. The simulation results in 9XX diesel engine

For 9XX diesel engine, the lubrication performance of the connecting rod bearing under two working conditions is evaluated. Figure 3 shows the peak oil film pressure (POFP) and the minimum oil film thickness (MOFT) plots of the connecting rod bearing in 9XX diesel engine. Figure 4 shows the peak asperity contact pressure (PASP) distribution diagram. The simulation results of 9XX diesel engine are listed in Table 2. It can be found that when the shaft rotational speed is set at 1405rpm, the maximum POFP and minimum MOFT of the connecting rod bearing under 80% load are 116.2Mpa and 0.77μm, respectively. The calculated film thickness ratio is 1.165. Therefore, the bearing runs in mixed lubrication. The slight asperity contact distributes in the range of 100°~120° of the upper connecting rod bearing. The maximum PASP is 18.4Mpa. While for the working conditions of 1455rpm and 100% load, the maximum POFP and the minimum MOFT are 139.8Mpa and 0.51μm, respectively. The film thickness ratio decreases to 0.772. The bearing lubrication state changes from mixed lubrication to boundary lubrication. The asperity contact still appears on two sides of the bearing. However, the degree of the asperity contact is more serious and the maximum PASP is increased to 44.6Mpa.

Table 2. Simulation results of 9XX diesel engine.

Parameter	1405rpm, 80% load	1455rpm, 100% load
Minimum MOFT (μm)	0.77	0.51
Maximum POFP (Mpa)	116.2	139.8
Maximum PASP (Mpa)	18.4	44.6
Film thickness ratio	1.165	0.772
Lubrication condition	Mixed lubrication	Boundary lubrication

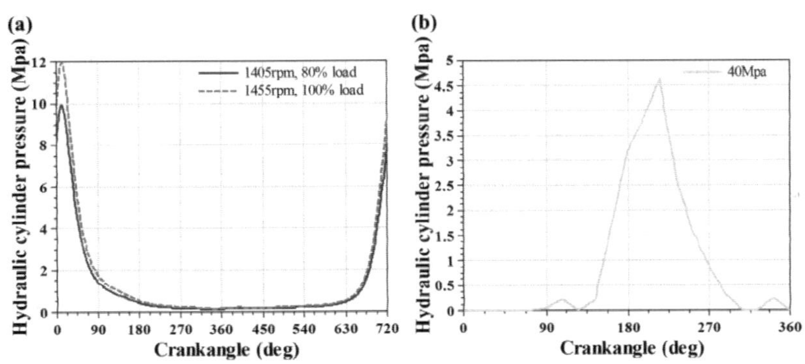

Figure 2. The plots of the hydraulic cylinder pressure: (a): 9XX diesel engine; (b): journal bearing fatigue test-rig at Glacier Vandervall.

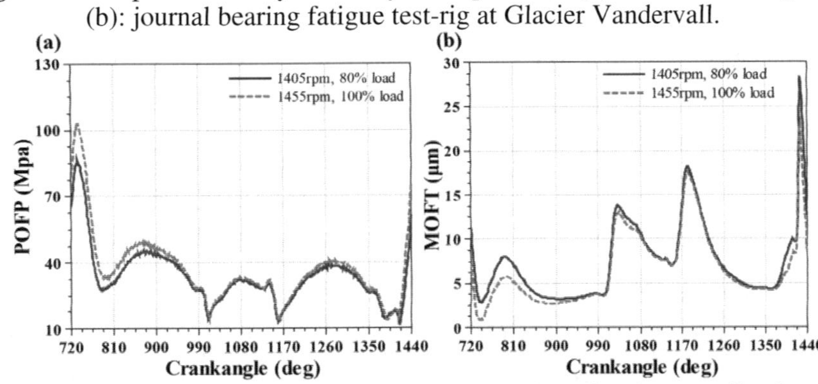

Figure 3. The simulation results of the connecting rod bearing in 9XX diesel engine: (a): the peak oil film pressure (POFP); (b): the minimum oil film thickness (MOFT).

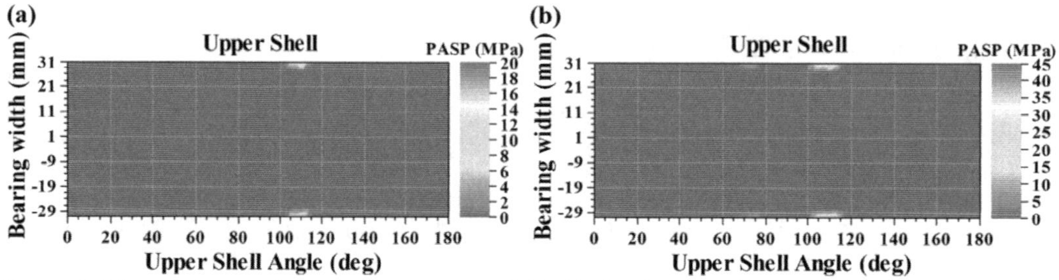

Figure 4. The peak asperity contact pressure (PASP) distribution diagram of the connecting rod bearing in 9XX diesel engine: (a): 1405rpm, 80% load; (b): 1455rpm, 100% load.

4.2. The simulation results in journal bearing fatigue test-rig

For journal bearing fatigue test-rig at Glacier Vandervall, the lubrication performance of the connecting rod bearing under three shaft rotational speeds (3500rpm, 4000rpm, and 4500rpm) and three relative clearances (1‰, 2‰, and 3‰) is discussed. The simulation results are summarized in Table 3.

According to the simulation results of Table 3, it can be found that at the same relative clearance, the maximum POFT and the minimum MOFT gradually increases with the increase of the speed. When the relative clearance is set at 1‰, the POFP and MOFT of the connecting rod bearing under different shaft rotational speeds are shown in Figure 5. The obtained film thickness ratios indicate that the bearing lubrication state is mixed lubrication and is not affected by the change of the speed. When the relative clearance is set to 2‰, the bearing lubrication state changes from boundary lubrication to mixed lubrication with the increase of the speed. The lubrication performance is improved. While for the relative clearance of 3‰, the increase of the speed does not affect the bearing lubrication state, which remains boundary lubrication. Therefore, in order to ensure the reliability and the service life of the connecting rod bearing in journal bearing fatigue test-rig, the speed should not be too low. In addition, when the relative clearance is 2‰, the bearing lubrication performance can be improved by increasing the speed.

While at the same speed, the maximum POFT gradually increases and the minimum MOFT gradually decreases with the increase of the relative clearance. Simultaneously, the bearing lubrication state changes from mixed lubrication to boundary lubrication. The lubrication performance is improved. When the speed is 3500rpm, the POFP and MOFT of the connecting rod bearing under different relative clearances are shown in Figure 6. Hence, it is necessary to choose an appropriate relative clearance to ensure better lubrication performance of the bearing. The relative clearance should not be too large.

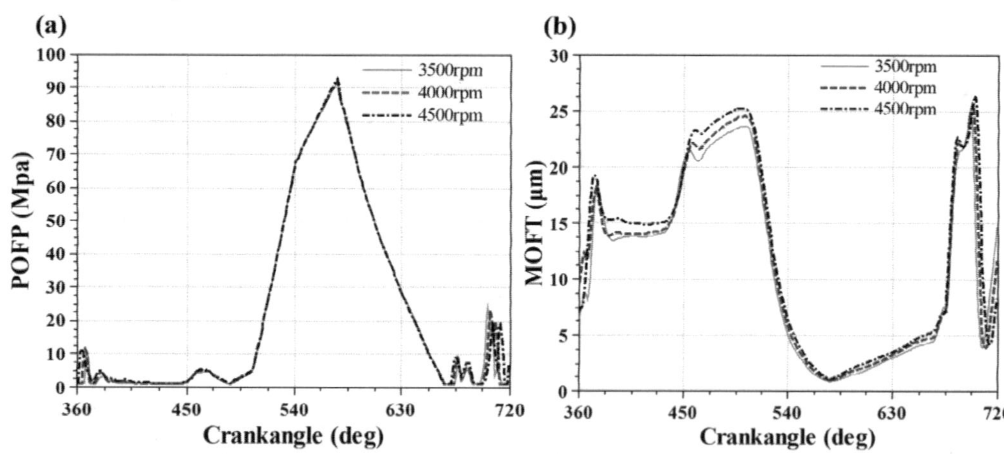

Figure 5. The simulation results of the connecting rod bearing in journal bearing fatigue test-rig at Glacier Vandervall under the relative clearance of 1‰ for different speeds:
(a): the peak oil film pressure (POFP); (b): the minimum oil film thickness (MOFT).

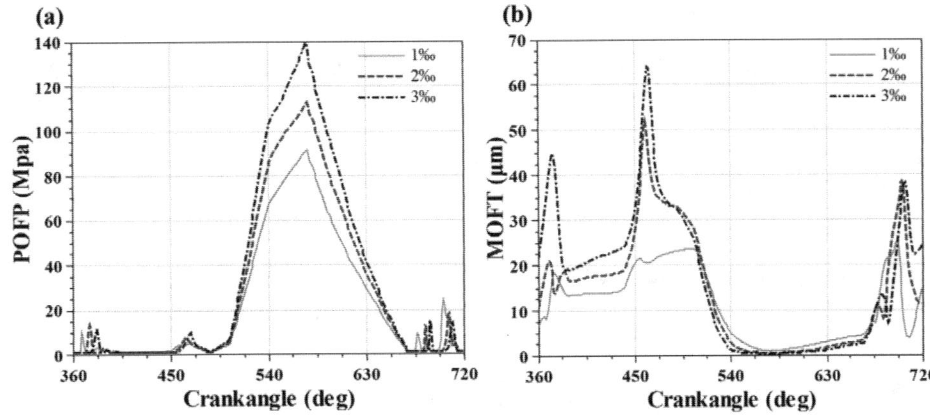

Figure 6. The simulation results of the connecting rod bearing in journal bearing fatigue test-rig at Glacier Vandervall at 3500rpm for different relative clearances: (a): the peak oil film pressure (POFP); (b): the minimum oil film thickness (MOFT).

Table 3. Simulation results of journal bearing fatigue test-rig at Glacier Vandervall.

Load (Mpa)	Shaft rotational speed (rpm)	Relative clearance	Minimum MOFT (μm)	Maximum POFP (Mpa)	Film thickness ratio	Lubrication condition
40	3500	1‰	0.83	91.4	1.687	Mixed lubrication
		2‰	0.47	112.6	0.955	Boundary lubrication
		3‰	0.30	139.8	0.610	Boundary lubrication
	4000	1‰	0.93	91.8	1.890	Mixed lubrication
		2‰	0.51	112.8	1.037	Mixed lubrication
		3‰	0.32	140	0.650	Boundary lubrication
	4500	1‰	1.02	92.5	2.073	Mixed lubrication
		2‰	0.54	113.5	1.098	Mixed lubrication
		3‰	0.33	140.3	0.671	Boundary lubrication

4.3. The comparison results
Based on the simulation results in Table 2 and Table 3, it can be found that for 9XX diesel engine, the minimum MOFT and the maximum POFT at 1405rpm and under 80% load are 0.77μm and 116.2Mpa, respectively. While for journal bearing fatigue test-rig, under the fixed load condition, the minimum MOFT and the maximum POFT at 4500rpm and a relative gap of 2‰ are 0.54μm and 113.5Mpa, respectively. The minimum MOFT values of two devices differ by 0.23μm and the maximum POFT values are close. Besides, the lubrication states of two devices are mixed lubrication. So it can be considered that the lubrication performance of two devices is close. Similarly, when the connecting rod bearing of 9XX diesel engine runs at 1455rpm and under 100% load, the minimum MOFT and the maximum POFT are 0.51μm and 139.8Mpa, respectively. While for journal bearing fatigue test-rig, under the fixed load condition, the minimum MOFT and the maximum POFT at 4500rpm and a relative gap of 3‰ are 0.33μm and 140.3Mpa, respectively. The minimum MOFT values of two devices differ by 0.18μm and the maximum POFT values are similar. The lubrication states of two devices are boundary lubrication. Therefore, under the corresponding working conditions of two devices, the lubrication performance is similar. The equivalent working conditions can provide guidance for the bearing lubrication performance evaluation in different devices.

5. Conclusion
In this paper, the lubrication performance of the connecting rod bearings in 9XX diesel engine and journal bearing fatigue test-rig at Glacier Vandervall is studied by AVL EXCITE. The main conclusions are as follows:

(1) For 9XX diesel engine, the connecting rod bearing runs in the mixed lubrication under 80% load when the speed is 1405rpm. The lubrication state changes to boundary lubrication when the connecting rod bearing runs at 1455rpm under 100% load. Under two working conditons, the asperity contact appears between the connecting rod bearing and the shaft, which is distributed in the range of 100°-120° upper shell angle.

(2) For journal bearing fatigue test-rig at Glacier Vandervall, under the same hydraulic cylinder pressure and relative clearance, the maximum POFT and the minimum MOFT gradually increases with the increase of the speed. When the relative clearance is set to 2‰, the bearing lubrication performance can be improved with the increase of the speed.

(3) For journal bearing fatigue test-rig at Glacier Vandervall, under the same hydraulic cylinder pressure and speed, the maximum POFT gradually increases and the minimum MOFT gradually decreases with the increase of the relative clearance. The bearing lubrication performance can be improved.

(4) The bearing lubrication performance of 9XX diesel engine at 1405rpm under 80% load are similar to that of journal bearing fatigue test-rig at Glacier Vandervall at 4500rpm and a relative clearance of 2‰ under a maximum cylinder pressure of 40Mpa. The lubrication states are mixed lubrication; While the bearing lubrication performance of 9XX diesel engine at 1455rpm under 100% load are similar to that of journal bearing fatigue test-rig at Glacier Vandervall at 4500rpm and a relative clearance of 3‰ under a maximum cylinder pressure of 40Mpa. The lubrication states are boundary lubrication.

(5) In future work, the influence of different surface roughness parameters, lubricating oil temperature, and viscosity on the lubrication performance of the connecting rod bearing will be studied in detail. Simultaneously, for 9XX diesel engine and journal bearing fatigue test-rig at Glacier Vandervall, the equivalent parameters corresponding to similar lubrication states, such as surface roughness, lubricating oil temperature, and viscosity will be calculated and compared.

Acknowledgment
The authors acknowledge the financial support provided by Marine Low Speed Engine Project-Phase I (Grant No. CDGC01-KT11).

References

[1] M. Nagata, M. Fujita, M. Yamada, T. Kitahara, Evaluation of tribological properties of bearing materials for marine diesel engines utilizing acoustic emission technique, Tribol. Int. 46 (1) (2012) 183–189.

[2] A. Vencl, A. Rac, Diesel engine crankshaft journal bearings failures: Case study, Eng. Fail. Anal. 44 (2014) 217–228.

[3] Mazhar Iqbal, Tribology: science of lubrication to reduce friction and wear, Int. J. Mech. Eng. Robot. Res. 2 (2014) 648-668.

[4] D. Y. Dhande, D. W. Pande, Multiphase flow analysis of hydrodynamic journal bearing using CFD coupled Fluid Structure Interaction considering cavitation. Journal of King Saud University - Engineering Sciences, 30 (2018) 345-354.

[5] D. Sfyris, A. Chasalevris, An exact analytical solution of the Reynolds equation for the finite journal bearing lubrication. Tribol. Int. 55 (2012) 46-58.

[6] H. Montazeri, Numerical analysis of hydrodynamic journal bearings lubricated with ferrofluid, Proceedings of the Institution of Mechanical Engineers, Part J: Journal of Engineering Tribology, 222 (2008) 51-60.

[7] K. P. Gertzos, P. G. Nikolakopoulos, C. A. Papadopoulos, CFD analysis of journal bearing hydrodynamic lubrication by Bingham lubricant, Tribol Int, 41 (2008) 1190-1204.

[8] B. Manshoor, M. Jaat, Z. Izzuddin, K. Amir, CFD Analysis of Thin Film Lubricated Journal Bearing, Procedia Engineering, 68 (2013) 56-62.

[9] Q. Lin, Z. Wei, N. Wang, W. Chen, Analysis on the lubrication performances of journal bearing system using computational fluid dynamics and fluid–structure interaction considering thermal influence and cavitation, Tribol Int, 64 (2013) 8-15.

[10] S K Guha, On the Steady-State Performance of Hydrodynamic Flexible Journal Bearings of Finite Width Lubricated By Ferro Fluids with Micro-Polar Effect, Int. J. Mech. Eng. Robot. Res. 1 (2012) 32-49.

[11] P. N, C. H, An average flow model for determining effects of three dimensional roughness on partial hydrodynamic lubrication, ASME, Trans J Lubric Technol 100 (1978) 12-17.

[10] V. H, The law of the relation between the viscosity of liquids and the temperature, Physica Z 22 (1921) 645-646.

[11] G. J, W. J, Contact of nominally flat surfaces., Proceedings of the Royal Society of London. Series A, Mathematical and Physical Sciences, 295 (1966) 300-319.

[12] B. J, Hamrock, Fundamentals of Fluid Film Lubrication, NASA Reference Publication, (1991).

Seismic performance and fragility functions of confined masonry old infrastructure with handmade bricks

G H Gonzales[1], A G Aguilar[1], G Huaco[2] and D Garber[3]

[1] Bachelor, Peruvian University of Applied Sciences UPC. Lima, Peru.
[2] Professor, Peruvian University of Applied Sciences UPC. Lima, Peru.
[3] Associate Professor, Florida International University FIU. Miami Florida, USA.

Email: u201521566@upc.edu.pe; u201517071@upc.edu.pe; pccighua@upc.edu.pe; dgarber@fiu.edu

Abstract. In several countries, confined masonry structures with handmade bricks were built for common to essential infrastructure. These bricks were made by clay using artisanal ovens; therefore, they have mechanical properties less than seismic codes currently recommend. Many of these buildings are one to three stories and were designed following past design codes or were built without considering any design codes. There is an uncertainty of the strength or ductility for these buildings to resist a severe earthquake event. For this purpose, this study aimed to develop fragility functions as way to estimate and quantify the vulnerability of these structures. This research describes a methodology to find fragility functions based on a non-linear model with three levels of masonry handmade bricks. Experimental test data was used to validate the proposed model during the calibration process. Incremental dynamic analyses were developed for 11 pairs of seismic records for both orthogonal directions. Fragility functions reported a high probability of collapse for demand levels from a design earthquake to maximum capable earthquake.

1. Introduction

In Peru and other developing countries on the coast of South America, the most common structural system of construction is confined masonry. According to the 2017 National Censuses [1], confined masonry represents almost 56% of the total infrastructure in Peru. Due to their age, many of these structures were built with handmade bricks and were designed before the release of the first Peruvian technical standard [2] in 1970 without considering factors such as material ductility, soil type, and displacement control of the wall's limits. Additionally, Jiménez and Moggiano [3] mentioned the sequence of twentieth and twenty-first century earthquakes: 19740 (Mw 9.0), 1966 (Mw 8.1), 1974 (Mw 8.1) and 2007 (Mw 8.1) has not released all the seismic accumulated after the occurrence of the grater 1746 Callao earthquake and tsunami (*Mw 9), rather only around 20%.

Faced with this catastrophic natural disaster, Simões et al. [4] is necessary to assess the capacity of these constructions to resist a high magnitude earthquake by developing analytical fragility functions that quantify the vulnerability through the probability of equaling or exceeding a certain damage limit state based on a seismic demand. In this paper, a way to assess seismic vulnerability of structural confined masonry system is presented. Eleven (11) pairs of seismic records were selected, which were treated for later use in the application of Incremental Damage Analysis (IDA) to a 60-year-old hospital structure with 3 stories of confined masonry with handmade bricks. From this, statistical processing of results was performed to develop fragility functions for each damage limit state, which will allow to evaluate the seismic vulnerability of the hospital under study.

2. Methodology

2.1. Description of structure analyzed

The structure studied is located in Lima, Peru. It is a hospital of confined masonry with handmade bricks that was built more than 60 years ago. It has a non-appropriate structuration to the actual NTE-030 with banked beams only one way and with insufficient joints (3cm) between adjoining buildings. Also, it has 3 stories with 3 meters of high each one. The handmade bricks had a lower axial compression (f'_m = 4.7 MPa) than masonry units used today. The mechanical properties of the materials were obtained from Gibu et al. [5]. Figure 1 shows the structure.

(a) (b)

Figure 1. (a) Analyzed infrastructure (b) Handmade bricks

2.2. Seismic records

According to FEMA [6], eleven pairs of seismic records were selected from stations located in the same soil type as the structure being analyzed. The soil type was rigid soil (S1) as defined by Ministry of Housing [7]. Also, these 11 pairs of seismic records were scaled to be compatible with Peruvian target spectra based on Al et al. [8].

2.3. Nonlinear model proposed

The nonlinear model of the structure uses the experimental database of real walls of masonry handmade bricks on a real scale by Huaco [9]. The equivalent nonlinear model proposed by Gonzales et al. [10] was used in this research. This model was validated by replicating the laboratory test from Huaco [9] in SAP 2000 where parameters of the pivot hysteresis were calculated based on Ilki and Kumbasar [11]. Then, the wall was subjected to a nonlinear time history analysis inducing the history of displacement of the tested wall to compare the hysterical responses obtained from SAP 2000 and the laboratory test. Finally, an iterative process was performed by varying the previously calculated pivot parameters of the hysterical curve to obtain a better match with the experimental model comparing hysteretic responses.

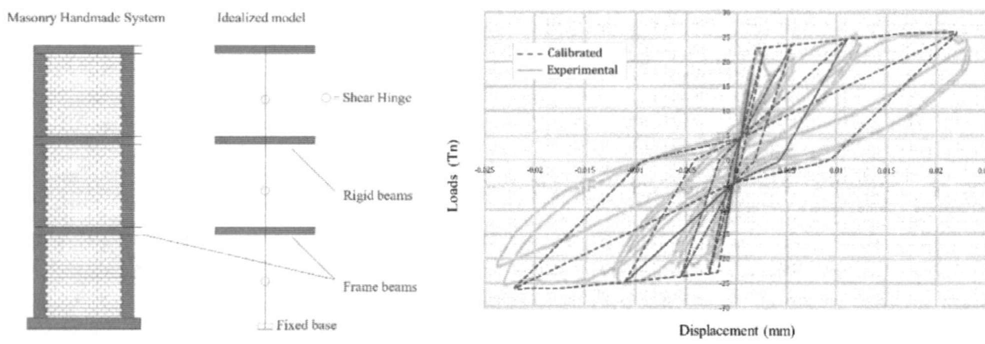

(a) (b)

Figure 2. (a) Shear wall modelling (b) Comparison of experimental and calibrated hysterical curve

2.4. Incremental dynamic analysis

The 11 pairs of seismic records were scaled to different peak ground acceleration (PGA) levels for each orthogonal direction, thus having 440 seismic records and the same number of nonlinear time history analyses. To plot the IDA curves, the intensity measure (IM) parameter was defined as the PGA and the damage measure (DM) parameter was defined as the interstory drift ratio. Figure 3 (a) and (b) show the family IDA curves highlighting 16%, 50% and 84% percentiles in each orthogonal direction.

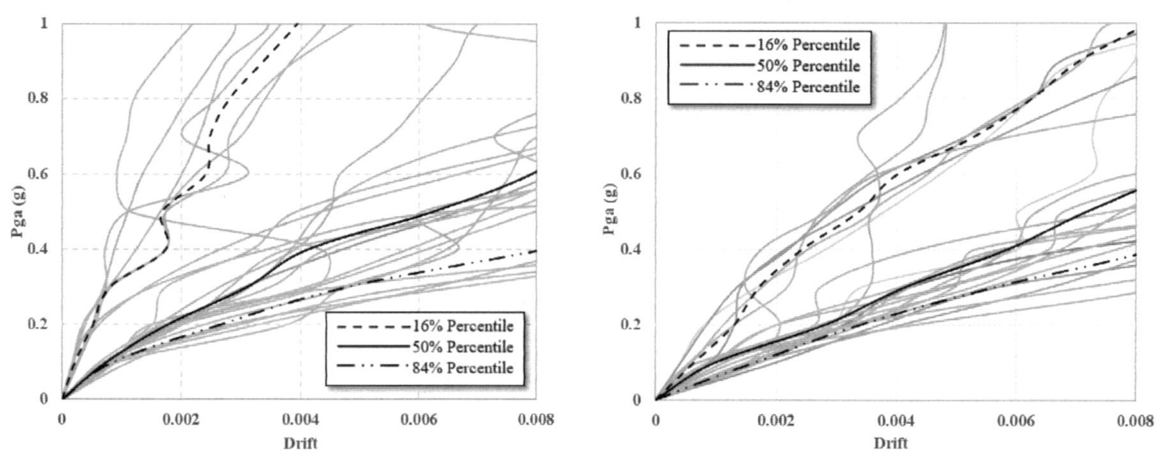

(a) (b)

Figure 3. (a) Percentiles 16%, 50% and 84% of the IDA curves in E-W direction (b) Percentiles 16%, 50% and 84% of the IDA curves in N-S direction.

Also, Figure 3(a) and (b) show for the same levels of demand the structure reported higher drift in N-S direction than in the E-W direction, which reflects the wall densities.

2.5. Generation of fragility Functions

2.5.1. Damage limit states. These are based on experimental test of real-scale walls for confined masonry with handmade bricks developed by Zavala et al. [12], where drifts are related to performance level: Operational (O) is 1/2500, Immediate Occupancy (IO) is 1/769, Life Safety (LS) is 1/350, Collapse Prevention (CP) is 1/286, Collapse (C) is 1/250.

2.5.2. Analysis statistical. Fragility functions were calculated with the maximum likelihood method. It mitigates non-convergence problems in adjusting fragility functions by making better predictions, which are better aligned with the damage observed in past seismic events [13].

IOP Conf. Series: Materials Science and Engineering **999** (2020) 012006 doi:10.1088/1757-899X/999/1/012006

2.5.3. Seismic demand estimation. Four levels of seismic demand were defined in this research, as indicated in ATC-40 [14], using the PGA for the Z4 zone indicated by the Ministry of Housing [7] as shown in Table 1.

Table 1. Seismic Demand according to ATC-40 associated to NTE-E 030

Seismic Demand	Description	PGA
The Serviceability Earthquake (SE)	Tr= 75 year	0.23g
The Design Earthquake (DE)	Tr= 500 years	0.45g
The Maximum Earthquake (ME)	Tr=1000 years	0.56g
The Maximum Capable Earthquake (MCE)	Tr=5000 years	0.68g

3. Results

Figure 4 (a) and (b) report the fragility functions for both orthogonal directions highlighting the probability of excess of the damage limit-states for each one demand level.

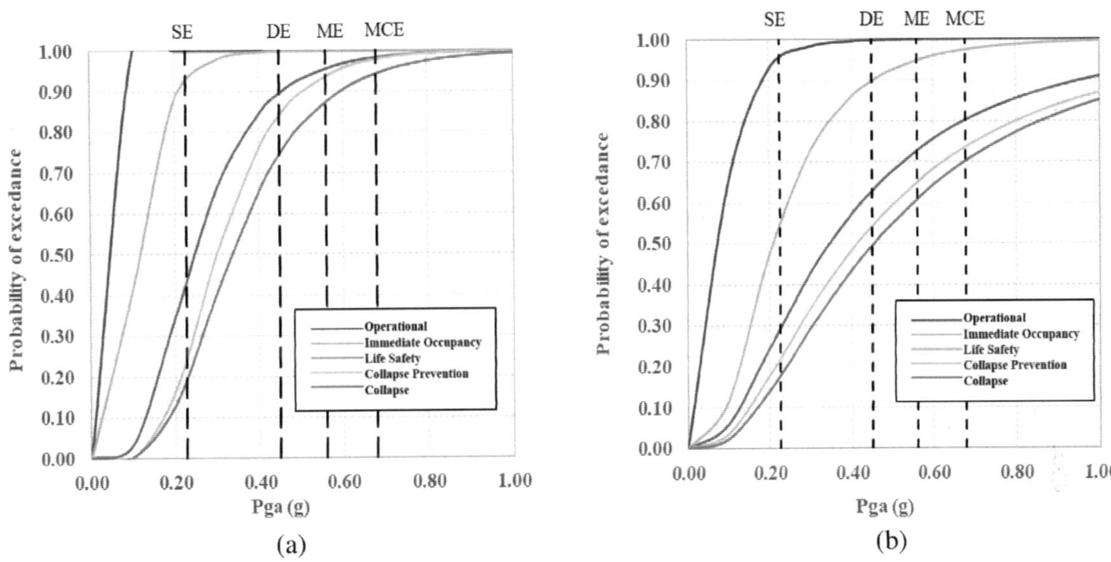

Figure 4. (a) Fragility functions in the N-S direction (b) Fragility functions in the E-W direction.

Table 2 and 3 shows probability to each damage limit state for different seismic demands. For instance, Table 3 shows for a design earthquake demand the probability of Collapse is 49.61%, the probability of Collapse Prevention is 4.68%, Life Safety is 8.64%, Immediate Occupancy is 26.76% and Operational 10%.

Table 2. Damage probability matrices of N-S

Seismic Demand	Operational	Immediate Occupancy	Life Safety	Collapse Prevention	Collapse
SE	6.95%	47.24%	19.24%	5.70%	20.87%
DE	0.23%	10.22%	5.32%	9.49%	74.75%
ME	0.05%	4.40%	1.78%	6.31%	87.45%
MCE	0.01%	1.80%	0.42%	3.52%	94.24%

Table 3. Damage probability matrices of E-W

Seismic Demand	Operational	Immediate Occupancy	Life Safety	Collapse Prevention	Collapse
SE	38.80%	26.14%	7.50%	4.22%	18.66%
DE	10.00%	26.76%	8.64%	4.68%	49.61%
ME	5.13%	22.03%	7.75%	4.07%	60.91%
MCE	2.58%	17.19%	6.60%	3.34%	70.26%

4. Conclusions

The results of the fragility functions show the high seismic vulnerability of the analyzed type of structure built by handmade bricks, reporting high probabilities of exceeding the limit states of immediate occupation, life safety, prevention of collapse and collapse for the different levels of demand.

The 50% percentile of the IDA curves show that in the E-W direction starting from a demand level of 0.45g, the structure would exceed the 0.5% drift limit indicated in NTP E030, while IDA curves in N-S direction this limit will be reached starting from a demand level of 0.35g. According to the damage limit states for those drift levels; the structure will already be in the collapse range, having presented considerable damage to the masonry wall with handmade bricks. These results show the need to develop policies to reinforce this type of building as the structure analyzed in this research is representative of many critical structures in Peru and other developing countries along the coast of South America.

References

[1] "National Censuses: XII Population, VII Housing and III Indigenous Communities, 2017. Indígenas," 2017.

[2] "Ministry of Housing, Construction and Sanitation, Earthquake Design Standard NTP-030 Peruvian Technical Norm. Lima, 1970."

[3] C. Jiménez and N. Moggiano, "Numerical simulation of the 1940 Lima-Peru earthquake and tsunami (Mw 8.0)," *J. Seismol.*, vol. 24, no. 1, pp. 89–99, 2020.

[4] A. G. Simões, R. Bento, S. Lagomarsino, S. Cattari, and P. B. Lourenço, "Fragility Functions for Tall URM Buildings around Early 20th Century in Lisbon, Part 2: Application to Different Classes of Buildings," *Int. J. Archit. Herit.*, vol. 0, no. 0, pp. 1–17, 2019.

[5] P.Gibu et al., "Seismic Vulnerability Study: Structural, Non-Structural and Functional of the José Casimiro Ulloa Emergency Hospital - Miraflores" no. 14, pp. 1–213, 2013.

[6] FEMA, "Seismic performance assessment of buildings. Volume 2-Implementation Guide," *Fema*, vol. 2, no. December, 2018.

[7] Ministry of Housing, Construction and Sanitation, Earthquake Design Standard NTP-030 Peruvian Technical Norm. Lima, p. 80, 2018. [8] L. Al, N. Abrahamson, and M. Eeri, "An Improved Method for Nonstationary Spectral Matching," vol. 26, no. 3, pp. 601–617, 2010.

[9] G. Huaco, "Behavior of a masonry building built with handmade brick against lateral loads", National University of Engineering UNI, Lima, Peru, 2005.

[10] G. Gonzales, A. Aguilar, and G. H. Pe, "Incremental Dynamic Analysis of 60 years old Hospital with on infill Masonry Handmade Bricks", XVIII LACCEI International Multi-Conference For Engineering, Education And Technology - The OAS Summit of Engineering for the Americas, pp. 29–31, July 2020.

[11] A. Ilki and N. Kumbasar, "Hysteresis Model for Reinforced Concrete Members," no. January, pp. 20–25, 2000.

[12] C. Zavala, M. Diaz, E. Flores, and L. Cardenas, "Damage Limit States for Confined Masonry Walls Based on Experimental Test," *Tecnia*, vol. 29, no. 2, pp. 135–141, 2019.

[13] G. Cremen and J. W. Baker, "Improving FEMA P-58 non-structural component fragility functions and loss predictions," *Bull. Earthq. Eng.*, vol. 17, no. 4, pp. 1941–1960, 2019.

[14] ATC-40, "Applied Tecnology Council - 40," *Seismic Evaluation and Retrofit of Concrete Building Volume 1*. pp. 1–334, 1996.

IOP Conf. Series: Materials Science and Engineering **999** (2020) 012007 doi:10.1088/1757-899X/999/1/012007

Cyclic Behavior of CFRP as Diagonal Ties and Anchors to Rehabilitate Severe Damaged Masonry Wall

G Huaco[1] and J Jirsa[2]

[1] Professor, Peruvian University of Applied Sciences UPC, Lima Peru
[2] Emeritus Professor, The University of Texas at Austin, USA
Email: pccighua@upc.edu.pe; jirsa@uts.cc.utexas.edu

Abstract. Seismic rehabilitation using Carbon Fiber Reinforcement Polymer CFRP is more often used worldwide. CFRP jacketing is the most common layout for columns and walls, however installing the CFRP sheet as diagonal ties increase shear capacity to walls instead of jacketing procedures, besides less CFRP materials is used reducing the cost to the retrofit work. It is know that CFRP sheets can be debonded from the concrete or masonry surface causing that the sheets will not develop its tensile capacity. Then CFRP anchors are introduced. It is presented the behavior of CFRP sheets as diagonal ties working even though sliding and CFRP anchors which attach the diagonal ties to concrete base and top of the masonry wall. The diagonal ties work at full tensile capacity even though the debonding and buckling presented under compression forces. CFRP anchors also provided a resistance capacity against sliding.

1. Introduction

Masonry walls are frequently damaged by natural disasters such as earthquakes. The pattern of damage commonly seen in walls is shear cracking and sliding between the wall foundation interfaces or between the courses of the wall. Sliding between courses occurs often when the masonry wall interface has no internal reinforcement. Retrofit using steel plates is effective increasing ductility however installation is difficult because bolt connections and welding [6], see Figure 1a. Therefore it is necessary to explore new procedures to rapid repair instead of replace a new wall. Carbon Fiber Reinforced Polymer CFRP sheet is used to provide tension capacity to structural members. However debonding avoids these sheets to develop their full tension capacity as Alcaino and Santa Maria reports at [2] as it can be seen in Figure 1b. Pampanin and Akgusel [3] improved the attachment installing CFRP U-patches as Figure 1c shows, however important drop of shear capacity of wall is presented because debonding of the CFRP materials.

(a)

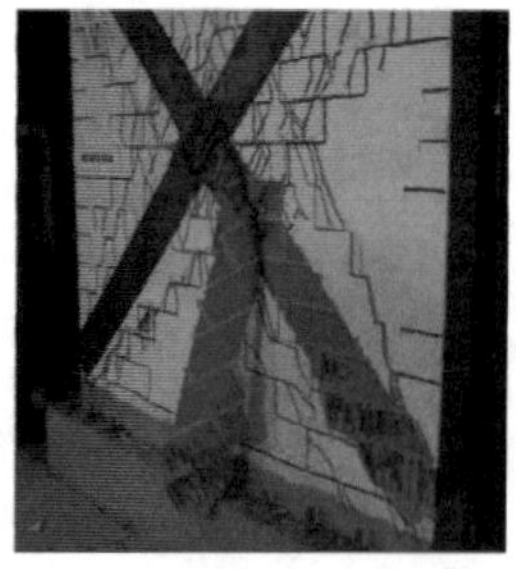
(b)

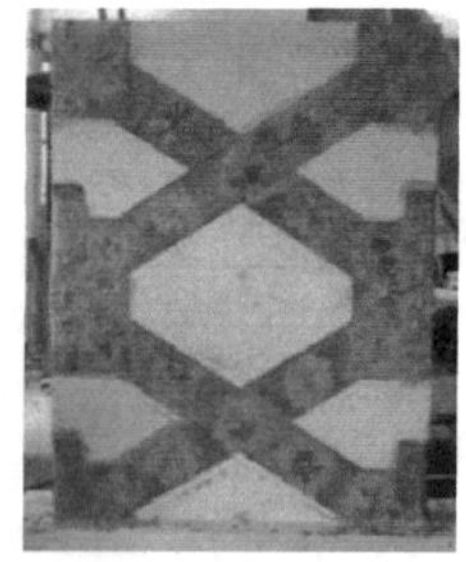
(c)

Figure 1. Masonry wall with steel plates [1] (a) with CFRP diagonal ties with no anchor system [2] (b) and including CFRP patches with no anchorage at corners [3] (c)

Hence it is necessary to apply CFRP anchors to resist shear and pulling forces to assure that CFRP sheets develop their full tensile deformation if debonding is presented [4, 5]. The objective this article aims to study the effectiveness of the CFRP materials to improve bothc shear capacity and provide ductilty to the masonry wall, in addition to reduce sliding at the top and bottom interfaces of the wall.

2. Rehabilitated Masonry Wall

A hollowed masonry wall was repaired after to receive severe damage such as sliding at the base, diagonal cracks and crushing of the toes at corners. The hollow concrete masonry units used to build the wall had normal dimensions16x8x8in and 1.25in wall thickness. The compression resistance was 3.1ksi. The wall was reinforced internally with #4 bars Grade 60 spaced vertically at 8in through the hollow cores in the concrete blocks and grouted. Dowels were installed. One #4 horizontal bar was placed between each course of the wall.

CFRP sheets of 0.04in thickness were attached to the wall along the diagonals to produce a tension brace or tie, and. CFRP anchors were installed to at the ends of the sheets. Figure 2a shows the location of the concrete ring and the CFRP materials. Then this was subjected to constant axial load of 160kips also 16 histeretic lateral cycles loads. Rotations at the top and bottom of the wall were restrained. Figure 2b shows the behavior of the repaired masonry wall. It is appreciated when occurs sliding and rupture of CFRP anchors. Details of test results are reported by Huaco & Jirsa [6].

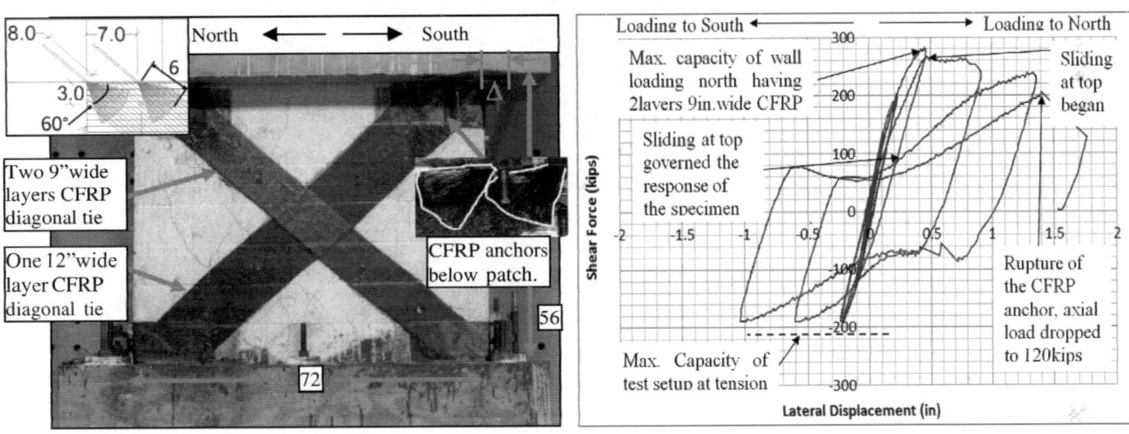

(a) (b)

Figure 2. Repaired masonry wall with CFRP ties and anchors, unit in inches (a) Hysteretic cycle response of the masonry wall [6]

3. Behavior of CFRP Materials

3.1. Response of the Two 9in. wide layers of CFRP Diagonal Tie Sheets

CFRP sheets worked as axial tension brace when the masonry wall was loaded to the south. Debonding of the CFRP diagonal strips appeared at the 5th cycle. This was noticed by a noise heard, being this more clear on the 9th cycle and further. In the 11th hysteretic cycle the maximum strain was reached in tension, 0.23% and 0.36% for the front and back face of the wall respectively. The pattern of the strain deformation is shown in Figure 3a and Figure 3b. Compression strains reached 0.20% (20% of the maximum tensile strain capacity). This was measured on the lower strain gage applied on the CFRP diagonal strip. The strain deformation in compression was larger for the bottom part of the CFRP diagonal strip, indicating that there was a major concentration of stress at the bottom tie. It is observed that the strain gages near the bottom exhibited larger strain than the gages near the top. The CFRP tie buckled as is shown in Figure 4.

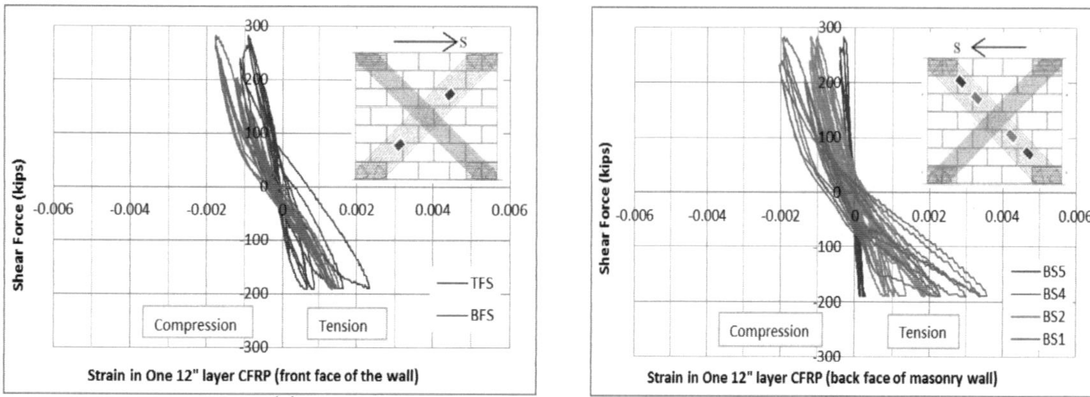

(a) (b)

Figure 3. (a) Strain deformation of the two 9in. layers CFRP sheet on west face of the masonry wall (b) Strain deformation of the two 9in. layers CFRP sheet on east face of the masonry wall.

Figure 4. Debonding and buckling of the CFRP tie of two 9in. layers

3.2. Response of the 12in.wide CFRP Diagonal Tie Sheets

CFRP sheets worked as axial tension punctual brace when the masonry wall was loaded to the north. In the 7th hysteretic cycle, some noise was heard indicating the beginning of the debonding. The CFRP anchors transferred tension from the CFRP sheets to the top and bottom support of the wall. The maximum strain measured at CFRP sheet at front face was 0.52% which is slightly above 50% of the strain capacity of the CFRP according to the producer's specification 1%. The maximum strain deformation measured on the CFRP sheet on the back face was 0.39%. Figure 1a and Figure 1b shows cycle behavior of the CFRP diagonal ties measured by strain gages.

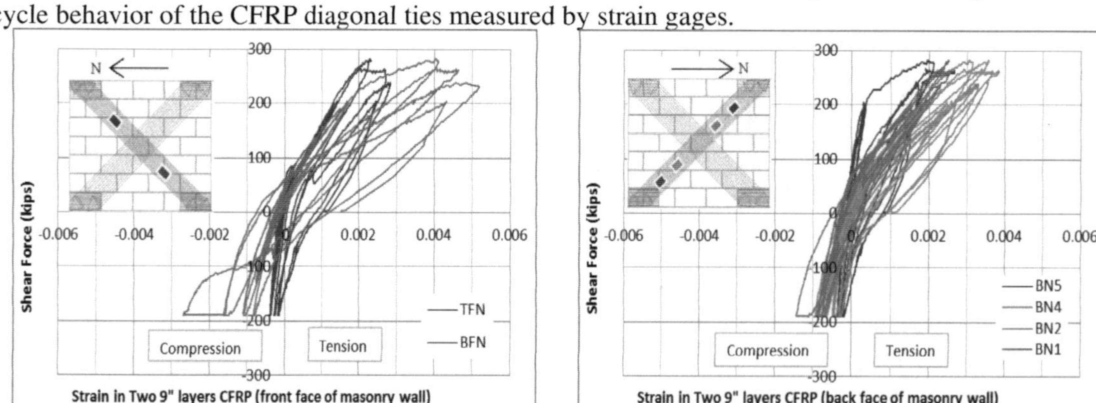

(a) (b)

Figure 5. (a) Strain deformation of the one 12 in. layer CFRP sheet on front face of the masonry wall (b) Strain deformation of the one 12 in. layer CFRP sheet on back face of the masonry wall

Compressive strains in the CFRP strips were measured under loading to the south. The maximum strain in compression prior to buckling was 0.16%. However, when the buckling occurred CFRP on front face of the wall, the compressive strain was 0.27% as Figure 5a shows, and 0.14% on the back face of the wall as it can be seen in Figure 5b. It was observed also that the strain measured was longer on the bottom part of the diagonal ties, having concentration of compression stress on this region. The CFRP strip buckled as is shown in Figure 6.

Figure 6. Debonding of the CFRP tie of one 12in layer

3.3. Behavior of CFRP Anchors on Rehabilitated Masonry Wall

It was observed during the test that the CFRP anchor provided resistance against the sliding at top of wall. CFRP anchors provided to the rehabilitated wall a capacity against sliding. Figure 7a and Figure 7b shows the tensioned and broken CFRP anchors and the final sliding. The anchor worked until their full tension capacity previous the rupture. No evidence of pulling out was found.

(a) (b)

Figure 7. Initial and final condition of the CFRP Anchors on top south corner presenting slip and rupture on front face (a) and back face of the retrofitted masonry wall (b)

Figure 8 shows a body diagram of the forces implied to the behavior of the CFRP anchors when the specimen was loading to north direction. There are two types of forces applied to the anchor: pulling loads due the tension load from the diagonal ties, and the sliding forces due the lateral load from the top concrete beam. Each of these forces is divided by the correspondent number of anchors. Using the parallel and perpendicular forces component respectively to the direction of the anchor, the CFRP anchor resists two types of load: pulling load (parallel to the anchor's direction) and shear force (perpendicular to the anchor's direction). It is not considered the dowel effect on the sliding. Then,

Equation 1 and Equation 2 show the developed formulas for the two types of forces applied in the CFRP anchor in base of the body diagram of forces for the north corner:

$$F\ tension\ tie/4 + (F\ sliding/8)cos\alpha = CFRP\ anchor\ Pulling\ Force \quad (1)$$

$$(F\ sliding/8)sin\alpha = CFRP\ anchor\ Shear\ Force \quad (2)$$

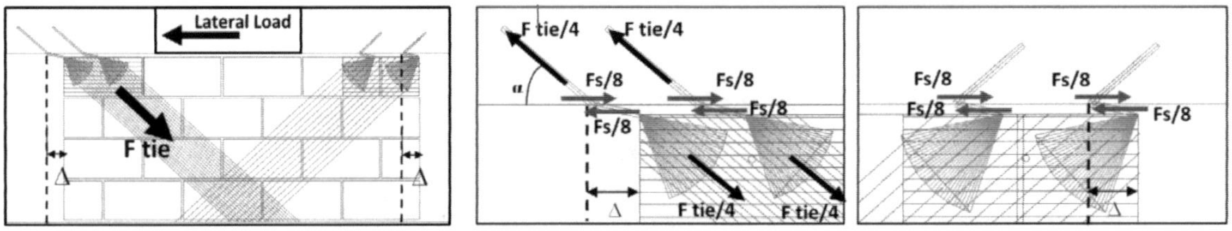

Figure 8. Forces applied to the CFRP anchor

Figure 8 also shows the body diagram for the south top corner of the masonry wall. It is considered that the CFRP diagonal ties had not tension pulling loads for this case. Therefore forces due sliding were considered only. It is deduced that the forces applied to the south CFRP anchors (or north CFRP anchors when it is loading to south) are due the sliding. Equation 3 and Equation 4 show the developed formulas by the body force diagram in order to calculate the forces applied to the CFRP anchor in the south corner.

$$F\ sliding/8\ cos\alpha = CFRP\ anchor\ Pulling\ Force \quad (3)$$

$$F\ sliding/8\ sin\alpha = CFRP\ anchor\ Shear\ Force \quad (4)$$

4. Conclusions

It is concluded that the CFRP anchors had a very important role for the resistance of the masonry wall against its sliding at top. Diagonal Ties also provided tension capacity to the retrofitted masonry wall.

CFRP ties regions at the opposite site of the sliding presents larger tension deformation than near regions and they also buckled, therefore it is very important the use of the CFRP anchor to assure that CFRP ties take complete tension forces, besides when compression occurs allow buckled ties to recover tension capacity.

For optimal use of CFRP materials, it is very important that strict quality control be carried out in the application of the CFRP. Without high quality installation, the capacity of the CFRP is compromised and may not reach the capacity desired.

It is recommended to develop a laboratory test program to study the response of CFRP anchor under cycle loads for pulling, beside shear forces at diagonal direction of the anchor simulating the action shown on this research.

References

[1] Taghdi M, Bruneau M and Saatcioglu M 2000 Seismic Retrofitting of Low-Rise Masonry and Concrete Walls using Steel Strips *J of Structural Engineering* pp. 1017-1025

[2] Alcaino P and Santa Maria H 2008 Shear Capacity of Masonry Walls Reinforced with Fibers *Proc. 33th South America Workshop of Structural Engineering* (Santiago, Chile)

[3] Pampanin S and Akgusel U 2011 Performance-Based Seismic Retrofit of Existing Reinforced Concrete Frame Building using Fiber-Reinforced Polymers: Challenges and Solutions *J Structural Engineering International SEI* vol 3 pp. 260-270

[4] Huaco G, Jirsa J and Bayrak O 2011 Quality control test for Carbon Fiber Reinforced Polymer (CFRP) anchors for rehabilitation *Proc. 10th Int. Symp. Fiber Reinforced Polymer Reinforcement for Concrete Structures SP-275 American Concrete Institute* (Tampa, Florida) pp. 675-713

[5] Akyuz O and Ozdemir G 2004 Mechanical Properties of CFRP Anchorages *Proc 13th World Conference on Earthquake Engineering* (Vancouver, Canada)

[6] Huaco G and Jirsa J 2017 Performance of Masonry Wall Retrofitted Using CFRP Sheets And Anchors *Proc 16th World Conference on Earthquake Engineering* (Santiago, Chile)

Numerical and experimental investigation of fully-coupled and uncoupled finite element model for electromagnetic forming of Aluminium Alloy Al 3014

Z Khan[1], M Khan[1], S H I Jaffery[1], M Younas[2], A Khan[3]

[1]Department of Design and Manufacturing Engineering (DME), School of Mechanical and Manufacturing Engineering (SMME), National University of Sciences and Technology (NUST), Sector H-12, 44000, Islamabad, Pakistan.
[2]Department of Mechanical Engineering, HITEC University, Taxila, 47080, Pakistan
[3] Department of Mechanical Engineering, Royal exchange building, University of Sheffield, UK
Email: zarak.khandme2015@smme.nust.edu.pk; mkhan@smm.nust.edu.pk; imran@smme.nust.edu.pk; Muhammad.younas@hitecuni.edu.pk; ashfaq.khan@sheffield.ac.uk

Abstract. Electromagnetic forming is a high-velocity forming process in which repulsive forces between conducting coil and workpiece are generated by mutual inductance, thus, creating opposite magnetic fields and Lorentz forces on conducting bodies. In such processes, the formability can be enhanced owing to the high-speed deformation of the material. This paper presents a fully-coupled numerical model and experimental study of the deformation behaviour of aluminium alloy (AL3014) using closed die magnetic pulse forming. The current density, magnetic flux density and Lorentz force of 1mm thick sheet were estimated numerically. The effect of deforming sheet on magnetic flux and induction was considered. As a result of which the percentage error in sheet deformation reduced from 10% for the existing uncoupled model to 5% for the fully coupled proposed model.

Keywords. Electromagnetic forming, Finite Element Model, Lorentz forces

1. Introduction

Electromagnetic forming involves the deformation of metal sheet under the influence of magnetic pressure. Due to high-speed deformation, the formability of material can be increased also the phenomena of wrinkling and spring back can be minimized as summarized by Psyk [1]. Experiments on free bulging of an annealed aluminium disk and its numerical simulation were first carried out by Takatsu et al [2]. Fenton and Daehn also worked on 2-D Arbitrary langrangian-Eulerian (ALE) to develop finite difference code (CALE) to predict the deformation of sheet metal [3] but two different time steps were used to maintain numerical stability of magnetic physics and material motion. A loose coupling method was used by Oliveira et al [4] to analyse the dynamic behaviour of sheet. ANSYS/EMAG and LS DYNA were used but the effect of change in geometry on the magnetic field wasn't considered. Corriea et al [5] used commercial finite element code ABAQUS/Explicit to study the deformation of sheet and influence of viscoplastic material behaviour in free bulge electroforming process neglecting the effect of velocity. Y.U Haiping et al[6] studied the effect of change in geometry on the magnetic force. The changing inductance of workpiece and hence circuit was ignored. Xiaohui et al [7] worked on adaptive remeshing technique in air mesh to change regularly as the sheet deforms.

IOP Conf. Series: Materials Science and Engineering **999** (2020) 012008 doi:10.1088/1757-899X/999/1/012008

Li et al[8] used a conducting material as a driver to deform a 0.5mm non-conducting titanium sheet. The model for the driver and the driven sheet was uncoupled. A fully coupled model for axisymmetric free bulging of aluminium alloy was developed by Quanliang Cao et al [9]. The effect of velocity on the current applied to the coil was considered. Simulated deformation of Cao gave much better results than previous uncoupled models. The model was for free bulging. Haiping Yu et al [10] discussed the comparison of conventional forming and electromagnetic forming. Circular hole flanging by electromagnetic forming showed better formability compared to conventional forming. The uncoupled numerical model was in good agreement with experimental results. Hak-Gon Noh[11] performed experiments and 3-D numerical simulations for sheet metal forming with unsymmetrical die. The uncoupled model showed satisfactory results.

The numerical model consists of an RLC circuit coupled with the electromagnetic and structural mechanics module. The proposed numerical model was used to validate the current densities, magnetic flux densities and Lorentz force during the forming process. Sheets were deformed experimentally and then validated numerically using the proposed numerical model. The morphology of experimentally deformed sheets was compared with numerically estimated results.

2. Numerical model

A Finite Element (FE) model was used to numerically solve the 2-Dimensional problem. In the simulation, the discharge current flowing through the coil was calculated by solving Eq.1. Magnetic flux and current density were calculated using Eq. (1 & 2). The magnetic force was then calculated using Eq. 4. The deformation of the workpiece was calculated in solid mechanics which use magnetic force as body load on the sheet. 2D geometry consisted of five domains as shown in Figure 1.

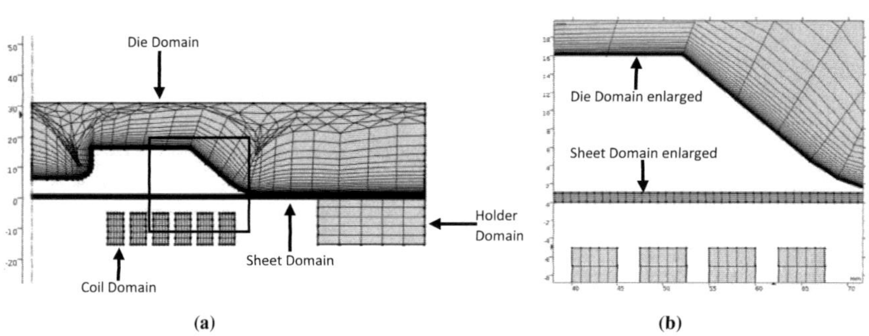

Figure 1. 2-D Mesh of closed die electromagnetic forming (a) Complete mesh (b) Sheet and die enlarged

The coil domain was connected with the current source that was generated in the electrical circuit module. The remaining four domains were sheet domain, die domain, air domain and blank holder domain. The magnetic field module was used to get the required Lorentz forces. The magnetic field module was fully coupled with the solid mechanics module to get the instantaneous deformation of the sheet with changing Lorentz forces with respect to time. A time-dependent study was used in the simulations.

$$I(t) = \frac{U0}{\omega L} e^{-\beta t} sin\,(\omega t) \qquad (1)$$

$$\nabla x\vec{E} = \frac{-d\vec{B}}{dt} \qquad (2)$$

$$\vec{J} = \frac{\vec{I_c}}{s} = \sigma_e \vec{E} \qquad (3)$$

Where $\vec{H}$ is magnetic intensity, $\vec{J}$: is current density, s: is a sectional area of one-turn of the coil, $\vec{B}$: magnetic flux density, $\vec{E}$: electric intensity, σ_e : electrical conductivity.

$$\rho \frac{d^2\overline{u}}{dt^2} - \Delta.\sigma s = \overline{fm} \qquad (4)$$

Where ρ is density, $\overline{\mathrm{u}}$ is the displacement vector, σs is stress tensor and $\overline{\mathrm{fm}}$ is the electromagnetic force density

The electrical and mechanical properties of the sheet and constants for the model were taken from the work of[11], where the constants used were $p = 6500s^{-1}$ & $m = 0.25$. For simulation of Sheet / AL3014. Cowper-Symonds model Eq.5 was used. Properties are given below in the Table.1

$$\bar{\sigma} = \sigma_y \left[1 + \left(\frac{\dot{\varepsilon}}{p}\right)^m\right] \quad (5)$$

Table 1. Electrical and mechanical properties of the sheet and coil [11]

Serial	Component/material	Properties	Parameter	Values
1	Forming Coil / Copper	Resistivity	ρ	1.72e-8 m
2	Sheet / AL3014	Resistivity	ρ	2.65e-8 m
		Poison's ratio	ν	0.35
		Density	ρ	2980 kg/m^3
		Elastic Modulus	E	69.0 GPa

3. Experimental details

After validation of the proposed model experiments were performed to compare the results of the model with experimental results. A schematic of the experimental setup is shown in Figure 2. The input parameters for the comparison were taken from the work of Noh at el., [11]. The input voltage used was 11kV, capacitance used was 333e-6 F and inductance was 5.8e-6H. Resistance and frequency were taken as 0.03[ohm] & 23876.10414[rad/s] respectively. The coil had an inner diameter of 50mm and an outer diameter of 140mm, its cross-sectional width was 5mm x10mm with a spacing of 2.5mm. Sheet blank was 260mm x 260mm with a thickness of 1mm. Experiments were performed at 20.146KJ energy.

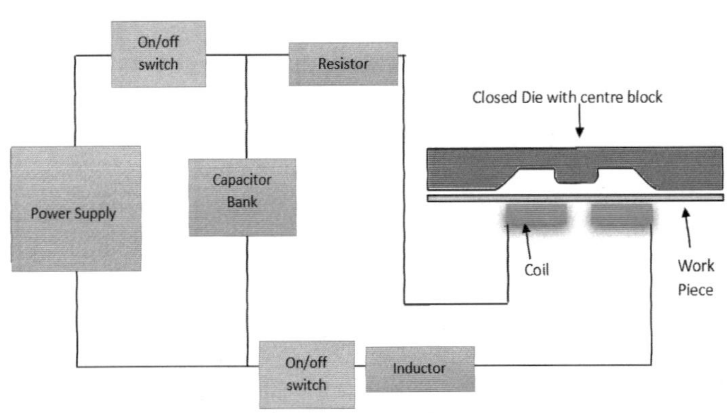

Figure 2. Electromagnetic forming configuration

4. Results

4.1. Comparison of current densities, magnetic flux densities and Lorentz force

The current densities, magnetic flux densities and Lorentz force at each time step were calculated numerically and plotted. The dotted curves represent the proposed fully coupled model while the solid curves are taken from reference article [11]. A comparison of the current densities at three known time steps is shown in Figure 3. The current density of the proposed model was observed to be conservative as compared to the uncoupled model. It was due to the effect of sheet deformation on the induction of the system and Lorentz force which was considered in the proposed fully coupled model and ignored in the uncoupled model developed by [11]. Similar results were observed for magnetic flux densities and Lorentz force as shown in Figure 4 and Figure 5 respectively.

The main reason was the effect of change in geometry on the current density & magnetic flux density in a fully coupled model which was ignored in the uncoupled model. The changing geometry changes the inductance of the workpiece and hence the circuit current which in turn changes the current density, magnetic flux density and Lorentz force [5]. The effect of temperature was ignored in the proposed model because the process takes place very quickly and the effect of temperature is always negligible [6]. From Table 2 it can be observed that the maximum value of current density at all time steps is lower for the proposed model as compared to the un-coupled model. The magnetic flux density estimated by the proposed model has lower peak value at 75μs as shown in Table 2, but has a slightly higher value at time step 170μs and 330μs. The Lorentz force at 75μs and 170μs has a lower value for the proposed model and slightly higher value at time step 330μs.

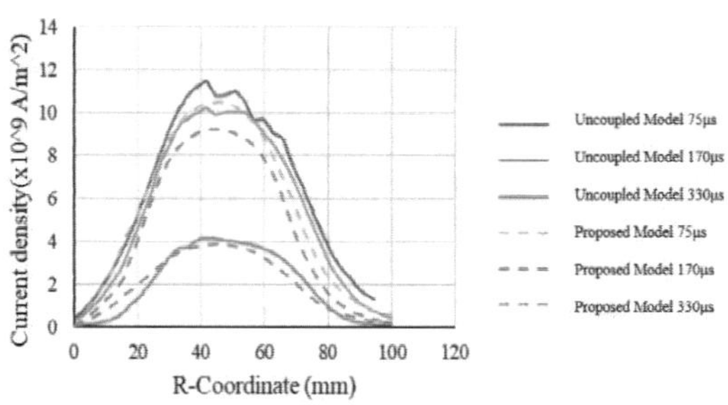

Figure 3. Comparison of Current densities results fully coupled proposed model vs Noh et al [11]

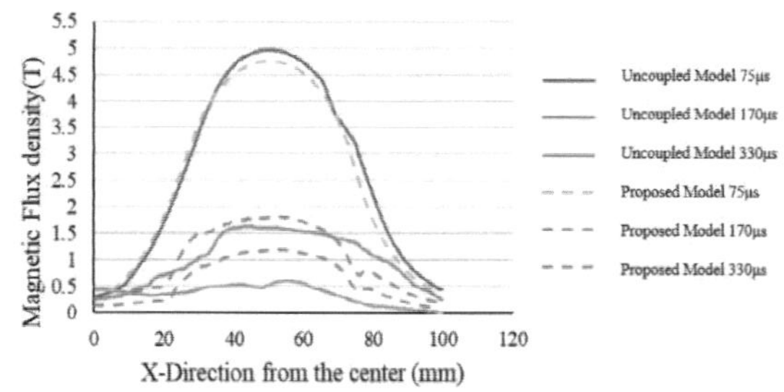

Figure 4. Comparison of Magnetic flux densities results fully coupled proposed model vs Noh et al [11]

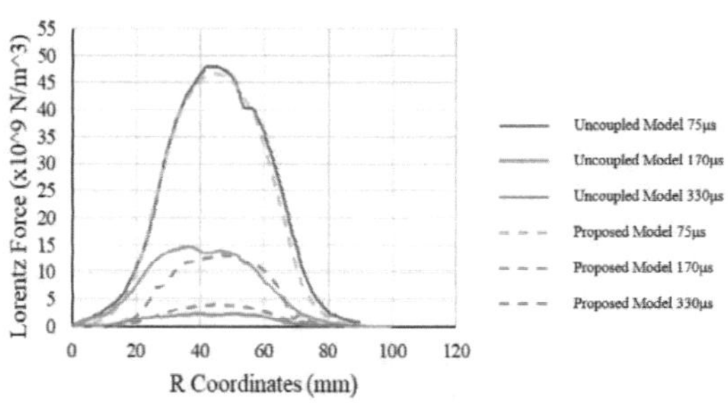

Figure 5. Comparison of Lorentz force results fully coupled proposed model vs Noh et al [11]

Table 2. Comparison of Peak values of all parameters at different time steps

Serial	Parameters	Time in [μs]	Uncoupled Model	Proposed Model	Difference %
1	Current Density	75	11.5×10^9 A/m2	10.5×10^9 A/m2	9%
		170	10×10^9 A/m2	9.4×10^9 A/m2	6.1%
		330	4×10^9 A/m2	3.8×10^9 A/m2	5%
2	Magnetic Flux Density	75	5T	4.8T	4.08%
		170	1.6T	1.8T	11%
		330	0.7T	1.1T	44%
3	Lorentz Force	75	4.8×10^{10} N/m3	4.65×10^{10} N/m3	3%
		170	1.4×10^{10} N/m3	1.3×10^{10} N/m3	7%
		330	0.3×10^{10} N/m3	0.4×10^{10}N/m3	28%

4.2. Z-Displacement of Deformed Sheet

From Figure 6 it can be observed that the numerical model used in the article [11] shows exaggerated results as compared to experimental values. On the other hand, with the proposed model much better results can be obtained by including the effect of changing geometry at each time step which results in a change in inductance and hence Lorentz force acting on the workpiece as it deforms. The maximum values of z displacement at three points were measured. The maximum z-displacment at point A for experimental, proposed model and uncoupled model was 12mm, 12.6mm and 13.3mm respectively. At point B the maximum values were 2mm,2.2mm and 2.4mm respectively. Similarly the maximum z-displacement at point C for experimental, proposed model and uncoupled model was 12.3mm, 12.8mm and 13.2mm respectively. The percentage error between the proposed model and experimental results at point A,B and C was 5%,10% and 4.06% respectively. while The percentage error between the uncoupled model and experimental results at point A.B and C was 10.83%,20% and 7.31%. The percentage error of the uncoupled model is approximately twice as that of the proposed model. The proposed fully coupled model will be used for further analysis.

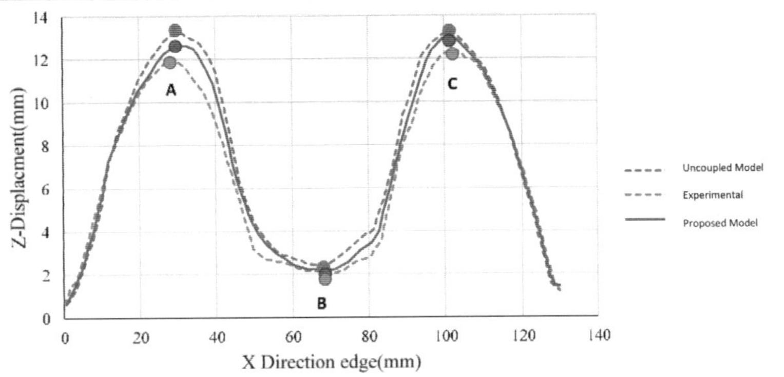

Figure 6. Comparison of Experimental Results with the simulated result

4.3 Comparison of experimental and numerical deformation

The energy of 20.146KJ was used to deform the sheet. After deforming the experimental profile for 1mm sheet was measured using FARO Arm CMM machine [12] and plotted against the numerically estimated curves as shown in Figure 7. A comparison of numerical and experimental curves at three points A,B and C was carried out. From Figure 7 it can be observed that the maximum displacement in z-direction on the numerical curve at point A_N ,B_N and C_N are 12.6mm,2.2mm and 12.7mm respectively. The maximum values in z-direction on experimental curves at A_E ,B_E and C_E are 12mm, 2mm and 12.1mm respectively. The percentage error between the experimental and numerical values was 4.7%, 10% and 5.7% respectively.

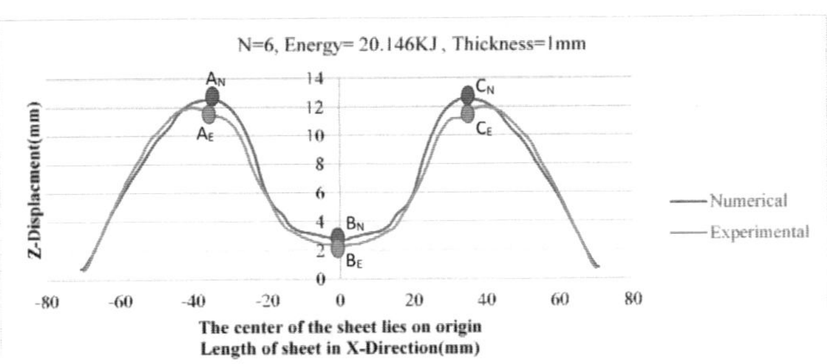

Figure 7. Comparison of Experimental results with the simulated result

Figure 8. Deformed sheet experiment at 20.146 KJ

It can be observed that numerical results are in close agreement with the experimental sheet deformation. The reason for the error is attributed to the minimal variation in the coil and sheet gap during experimentation, which is a critical factor in defining magnetic pressure [13]. Figure 8 shows the experimentally deformed sheet.

5. Conclusion

- The effect of sheet deformation on induced current and Lorentz force can result in deviation of numerical analysis from actual results during electromagnetic forming. From the results, it can be seen that the current densities estimated by the uncoupled model differ from the results obtained from the fully-coupled model with a maximum and minimum difference of 9% and 5% respectively.
- The magnetic flux densities estimated by the proposed model have a minimum difference of 4.08% at 75 μs. The difference then grows bigger till 330 μs.
- The peak Lorentz force estimated by the proposed model and uncoupled model are nearly the same with only 3% difference. The maximum difference is at the time step 330 of about 28%.
- The maximum percentage error between the uncoupled model and experimental values was 10.83% at point A and 20% at point B. Whereas the percentage error between experimental values and the proposed model was 5% at point A and 10% at point B. Although both the model gives exaggerated values, yet the proposed fully coupled model is closer to the experimental curve hence more accurate.
- The experimental validation shows the proposed model gives improved results as compared with the uncoupled model.

References

[1] V. Psyk, D. Risch, B. L. Kinsey, A. E. Tekkaya, and M. Kleiner, May 2011 Electromagnetic forming - A review, *J. Mater. Process. Technol.*, **vol. 211**, no. 5, pp. 787–829

[2] N. TAKATSU, M. KATO, K. SATO, and T. TOBE, 1988 High-speed forming of metal sheets by electromagnetic force., *JSME Int. journal. Ser. 3, Vib. Control Eng. Eng. Ind.*, **vol. 31**, no. 1, pp. 142–148

[3] G. K. Fenton and G. S. Daehn, 1998 Modeling of electromagnetically formed sheet metal, *J. Mater. Process. Technol.*, **vol. 75**, no. 1–3, pp. 6–16

[4] D. A. Oliveira, M. J. Worswick, M. Finn, and D. Newman, Dec. 2005 Electromagnetic forming of aluminum alloy sheet: Free-form and cavity fill experiments and model, *J. Mater. Process. Technol.*, **vol. 170**, no. 1–2, pp. 350–362

[5] J. P. M. Correia, M. A. Siddiqui, S. Ahzi, S. Belouettar, and R. Davies, Oct. 2008 A simple model to simulate electromagnetic sheet free bulging process, *Int. J. Mech. Sci.*, **vol. 50**, no. 10–11, pp. 1466–1475

[6] Y. U. Haiping, L. I. Chunfeng, and D. E. N. G. Jianghua, Jan. 2009 Sequential coupling simulation for electromagnetic-mechanical tube compression by finite element analysis, *J. Mater. Process. Technol.*, **vol. 209**, no. 2, pp. 707–713

[7] X. Cui, J. Mo, S. Xiao, E. Du, and J. Zhao, Numerical simulation of electromagnetic sheet bulging based on FEM.

[8] F. Li, J. Mo, H. Zhou, and Y. Fang, 3D Numerical simulation method of electromagnetic forming for low conductive metals with a driver.

[9] Q. Cao *et al.*, 2014 Dynamic analysis of electromagnetic sheet metal forming process using finite element method, *Int. J. Adv. Manuf. Technol.*, **vol. 74**, no. 1–4, pp. 361–368

[10] H. Yu, Q. Zheng, S. Wang, and Y. Wang, Jul. 2018 The deformation mechanism of circular hole flanging by magnetic pulse forming, *J. Mater. Process. Technol.*, **vol. 257**, pp. 54–64

[11] H. G. Noh, W. J. Song, B. S. Kang, and J. Kim, 2014 3-D numerical analysis and design of electro-magnetic forming process with middle block die, *Int. J. Precis. Eng. Manuf.*, **vol. 15**, no. 5, pp. 855–865

[12] FaroArm® - Portable 3D Measurement Arm for any application. [Online]. Available:

https://www.faro.com/products/3d-manufacturing/faroarm/. [Accessed: 12-May-2020].

[13] W. Xiong, W. Wang, M. Wan, and X. Li, Jul. 2015 Geometric issues in V-bending electromagnetic forming process of 2024-T3 aluminum alloy, *J. Manuf. Process.*, **vol. 19**, pp. 171–182

IOP Conf. Series: Materials Science and Engineering **999** (2020) 012009 doi:10.1088/1757-899X/999/1/012009

Thermomechanical and morphological properties of sustainable mortars employing blast furnace slag and fly ash reinforced cement

I Farina[1], R Singh[2,3], M Singh[2], P Preet[2], R Kumar[4], F Fraternali[5] and F Colangelo[1]

[1] Dept. of Engineering, University of Naples (Italy)
[2]Dept. of Production Engineering, Guru Nanak Dev Eng. College, Ludhiana (India)
[3]Dept. of Mechanical Engineering, National Institute of Technical Teachers Training and Research, Chandigarh (India)
[4]University Centre for Research and Development, Chandigarh University, Mohali (India)
[5]Dept. of Civil Engineering, University of Salerno (Italy)
Email: ilenia.farina@uniparthenope.it; rupindersingh78@yahoo.com; ms.hps1@gmail.com; pawankumar200708@gmail.com; ranvijayk12@gmail.com; f.fraternali@unisa.it; francesco.colangelo@uniparthenope.it

Abstract. The blast furnace slag (BFS) is non-metallic co-product (such as silicates and alumina silicates etc.) which absorbs sulphur from the charge and comprises of around 20% (by weight) of Fe production and its use as reinforcement in cement mortar has been widely explored to improve thermal and compressive properties along with addressing the sustainability issues related to Fe production. But hitherto little has been reported on effect of BFS and fly ash (FA) reinforced cement mortar from surface topography, thermal stability and morphological properties view point. This paper reports the comparison of compressive, thermal and morphological properties of cement mortar with air cooled BFS and FA reinforced cement mortar as a case study. The results of study suggests that 80% of BFS and 20% FA reinforced cement mortar (cement: (BFS+FA): 1: (0.8+0.2) by weight%) possess lowest thermal conductivity (0.65W/mK), greatest porosity (29.65%) and acceptable compressive strength (6.6MPa) in comparison to cement mortar comprising of cement: sand as 1:1. The results are supported with compressive strength data, optical photo micrographs, thermal analysis based upon differential scanning calorimetry (DSC), surface topography (based upon 3D rendered images).

1. Introduction

Recycling of industrial and agricultural solid waste is one of the suitable solutions for minimizing pollution as well as economic issues [1]. Waste materials such as FA, fuel waste, fluxing waste and plastic can be used for the production of bricks to reduce the pollutants [2] and for the production of sustainable materials [3]. To safeguard the environment, recycling of solid waste plays a vital role. Presently around 960 million Tons of waste is present in terms of industrial, agricultural and mining waste and re-use of these waste substances can help significantly to reduce the environmental pollution [4-7]. The use of these types of waste in civil construction may reduce the weight of concrete structures. It has been reported that thermal conductivity (k) is lower in light weight concrete in comparison to normal weight concrete which results in reduction of heat transfer and energy consumption [8]. Some studies have reported application of wool fibers in cement mortar panels and investigations for mechanical strength and insulation properties were made [9]. It has been observed that the use of wheat and barley straw as fiber reinforcement for unfired bricks resulted into decreases in thermal conductivity, while it increased with use of higher cement and gypsum content [10]. Application of FA and bottom ash separately in different proportions resulted into reduction of 'k' value of building material, thus increasing the energy efficiency and saving the natural mineral sand [11-12].

Use of boron active belite cement mortar with addition of FA in different combinations/proportions increased the compressive strength of mortar at high temperature and significantly improved the high temperature resistance [13]. To study the effect of FA and lime on transport performance and mechanical properties of cement mortar and its application in filling material to improve the strength of filling body has been explored by some researchers [14]. In order to understand the impact of fineness and replacement ratio of FA in different combinations resulting in increase of compressive strength and lowering water absorption and decrease of porosity in blended cement mortar, studies have been widely reported [15-16]. Also significant studies have been reported on application of recycled waste glass powder of ground granulated BFS or FA cement binder blend to investigate micro-structural properties which shows improved compressive strength with lower total porosity and exhibit a dense microstructure [17]. Some studies have reported that morphology of the matrix is more relevant to thermal conductivity of mortars as compared to chemical composition of its components. As regards to strength of mortars containing ground granulated BFS along with Portland cement reveals that mortars attain strength fast at high temperature and curing temperature significantly affect the ultimate strength [18]. In case of low and high exposure temperatures (i.e. up to 250 and 500 °C or greater), BFS can effectively improve the mortar's performance. The literature review reveals that significant studies have been reported on use of BFS and FA as reinforcement in cement. But hitherto little has been reported on effect of BFS and FA reinforced cement mortar from surface topography, thermal stability and morphological properties view point. This paper reports the comparison of compressive, thermal and morphological properties of cement mortar with air cooled BFS and FA reinforced cement mortar as a case study. The BFS has been procured from local industry in Naples (Italy) and cast of different proportions were prepared (11×11×3cm) after blending with cement.

2. Experimentation

Blast furnace slag and marble sludge were used as sand replacement at different percentages (0%, 40%+60%, 60%+40%, 80%+20%) by weight while fly ash were used as 20% (by weight) cement replacement (table 1). Samples having dimension of 11×11×3 cm were casted and cured for 28 days at room temperature.

Table 1. Material composition/ proportion of BF, MS and FA (by wt.%) with cement mortar.

No.	Nomenclature	Material composition	Material composition in ratio's
1	REF 1 REF 2	Cement mortar using 100% sand as aggregate	C:S 1:1
2	BFS40-1 BFS40-2	Cement mortar using 20% FA (by wt.%) as replacement of cement and 40% BFS and 60% MS (by wt.%) as replacement of sand	C:(BFS+MS) 1:(0.4+0.6)
3	BFS60-1 BFS60-2	Cement mortar using 20% FA (by wt.%) as replacement of cement and 60% BFS and 40% MS (by wt.%) as replacement of sand	C:(BFS+FA) 1:(0.6+0.4)
4	BFS80-1 BFS80-2	Cement mortar using 20% FA (by wt.%) as replacement of cement and 80% BFS and 20% MS (by wt.%) as replacement of sand	C:(BFS+FA) 1:(0.8+0.2)

3. Results and discussion

3.1 Thermal conductivity

In order to determine the thermal conductivity of samples, a thermal conductivity apparatus (of insulating slabs) was used.

The thermal conductivity of the tested samples was calculated after 10-minute intervals, when the thermal stability conditions were attained, and under fixed atmosphere conditions (results in table 2). Electric current was provided at frequency of 50 Hz, with phase difference (angle between current and voltage) as $\pi/4$. Voltage and current were provided to a heat source as 150V and 3.5A respectively. The average thermal conductivities k of the examined materials is reported in table 3. Based upon the results presented in table 2, one observes that the BF 80 mortar offers the best performances from the point of view of thermal insulation (minimum thermal conductivity). Such results are explained by the fact that the BFS80 mortar comprises a greater proportion of silicates/ alumina silicates, as compared to BFS40 and BFS60 mortars, which results in more controlled thermal conductivity.

Table 2. Observations for thermal conductivity of the examined samples (T1: temperature of surface attached with heat source, T2: temperature of upper surface, dT: difference in temperature, dx: sample thickness in m).

Samples	T1 (°C)	T2 (°C)	dT	dx	Thermal conductivity (W/mK)
REF-1	285	108	-177	0.0345	1.07
REF-2	310	88	-222	0.0347	0.95
BFS40-1	287	116	-171	0.0359	0.85
BFS40-2	245	73	-172	0.036	0.84
BFS60-1	272	138	-134	0.0335	0.69
BFS60-2	335	90	-245	0.0345	0.70
BFS80-1	394	89	-305	0.0306	0.66
BFS80-2	385	166	-219	0.031	0.65

Table 3. Observations of average thermal conductivity of two samples.

Samples	Average of thermal conductivity (W/mK)
REF	1.01
BFS40	0.85
BFS60	0.70
BFS80	0.65

3.2 Mechanical strength (Compressive strength)

The compressive strength of the mortar samples under examination was tested according to ASTM C39/C39M standard, by determining the peak load in kN for 4×4cm2 cross section specimens cut from the 11cm×11cm×3cm samples. A compressive testing machine with a capacity of 2000kN was employed, with an arrangement for automatic pacing. The experimental peak loads of the tested samples are shown in
Table .

Table 4. Observed average peak load and compressive strength.

Samples	Average peak load (kN)	Compressive strength (MPa)
REF	19.2	12
BFS40	16.04	10.025
BFS60	14.55	9.09
BFS80	10.57	6.6

The results presented in table 4 show that the compressive strength of REF mortar was the greatest, followed by that of BFS40, BFS60 and BFS80 mortars. In order to further interpret these behaviors, surface porosity tests were performed.

3.3 Morphology and porosity (percentage)
A tool maker microscope (Make: Mitutoyo, Japan) was used to reveal the morphologies of composite at ×30, while metallurgical image analyzer (MIAS) software was used for the analysis of material porosity at ×100. Table 5 shows the average porosities measured for the examined materials.

Table 5. Observations for average porosity (%).

Samples	Average porosity of samples (%)
REF	21.22
BFS40	18.51
BFS60	23.04
BFS80	29.65

4. Conclusions
Following are the conclusions from this study:

1. The porosity of samples of BFS80 was observed as 29.65% greater than the REF samples (21.22%). Hence, these samples are better for construction sites where there will be more water logging problems.
2. As regards to compressive strength of samples is concerned REF samples has shown maximum (12MPa), followed by BFS60 (9.09MPa), BFS40 (10.025MPa) and minimum was observed for BFS80 (6.6MPa).
3. Based upon observations of thermal conductivity BFS 80 resulted into thermal conductivity of 0.65W/mK even better than the REF samples (1.01W/mK), making this sample best from thermal conductivity and heat capacity view point.

In future work, we will study the design and development of innovative sustainable composites, which mix the materials studied in the previous works with different green materials [19-23]. Additional future research lines will investigate the use of eco-sustainable materials for the design of next-generation meta-materials and structures, whose properties mainly derive from the geometry of the microstructure [24-30].

Acknowledgement
The authors are thankful to Manufacturing Research Lab (Department of Production Engineering), Guru Nanak Dev Engineering College, Ludhiana, for providing research facilities.

References

[1] Ralegaonkar RV and Mandavgane SA 2011 *Constr. Build Mater.* **25**4037-42.
[2] Dondi M, Marsigli M and Fabbri B 1997 *Tile & Brick International* **13**(3) 218-25.
[3] Singh, R, Kumar R, Farina I, Colangelo F, Feo L and Fraternali F 2019 *Polym J.* **11**(1), 62.
[4] Coppola L, Bellezze T, Belli A, Bignozzi MC, Bolzoni F, Brenna, Carsana M, Colangelo F – Part 2 2018 *J Appl BiomaterFunc.* **16**(4), 207-21.
[5] Singh R, Singh H, Farina I, Colangelo F and Fraternali F 2019 *Compos Part B-Eng***156**, 259-65.
[6] Colangelo F, Petrillo A, Cioffi R, Borrelli C and Forcina A 2018 *Sci Total Environ*, **615** 1506-17.
[7] Colangelo F, Forcina A, Farina I and Petrillo A 2018 *Buildings* **8**(5) 70.
[8] Cardinale T, Arleo G, Bernardo F, Feo A and De Fazio P 2017 In 50th AiCARR International Congress on Beyond NZEB Buildings.

[9] Ashour T, Korjenic A, Korjenic S and Wu W 2015 *Energy Buildings* **104**139-46.
[10] Gökçe HS 2019 *J Clean Prod.* **211**992-1000.
[11] Roviello G, Menna C, Tarallo O, Ricciotti L, Ferone C, Colangelo F and Cioffi R 2015 *Mater Des* **87**, 82-94.
[12] Colangelo F, Cioffi R, Roviello G, Capasso I, Caputo D, Aprea P and Ferone C 2017 Compos Part B-Eng **129**, 11-17.
[13] Yan B, Kouame KJ, Lv W, Yang P and Cai M *J Mater Res Technol.* **8**(1) 1451-60.
[14] Hsu S, Chi M and Huang R 2018 *Constr. Build Mater.* **176**250-8.
[15] Roviello G, Ricciotti L, Tarallo O, Ferone C, Colangelo F, Roviello V and Cioffi R 2016 *Materials* **9** (6), 461.
[16] Mendes JC, Barreto RR, de Paula AC, da Fonseca Elói FP, Brigolini GJ and Peixoto RA 2019 *Cement Concrete Comp.* **104**103365.
[17] Shariq M, Prasad J and Ahuja AK 2008 *Asian J Civil Eng.* **9** 61-74.
[18] Lübeck A, Gastaldini AL, Barin DS and Siqueira HC 2018 *Cement Concrete Comp.* **34**(3) 392-9.
[19] Fraternali F, Farina I and Carpentieri G 2014 *Mech Res Commun.* **56** 18-25.
[20] Fabbrocino F, Farina I and Modano M 2017 *Compos Part B-Eng***115** 376-83.
[21] Singh R, Kumar R, Ranjan N, Penna R and Fraternali F 2018 *Compos. Struct.* **184** 704-13.
[22] Fabbrocino F, Farina I, Amendola A, Feo L and Fraternali F 2016 ECCOMAS Congress 2016 – Proceedings of the 7th European Congress on Computational Methods in Applied Sciences and Engineering, 1, 1893-1908.
[23] Kumar R, Singh R, Hui D, Feo L and Fraternali F 2018 *Compos Part B-Eng.* **134**, 193-06.
[24] Fraternali F and Feo L 2000 *Compos Part B-Eng.* **31** 141-58.
[25] Fortunato A, Fraternali F and Angelillo A 2014 *Ing Sismica-Ital.* **31**(1) 41-51
[26] Modano M, Fabbrocino, F, Gesualdo A, Matrone G, Farina I and Fraternali F 2015 COMPDYN 2015—5th ECCOMAS Thematic Conference on Computational Methods in Structural Dynamics and Earthquake Engineering 209-17.
[27] Fraternali F, Spadea S and Ascione L 2013 *Compos Struct.* **100** 280-89.
[28] Fabbrocino F, Amendola A, Benzoni G and Fraternali, F 2016 *Ing Sismica-Ital.* **1-2**, 62-71.
[29] Mascolo I, Amendola A, Zuccaro G, Feo L and Fraternali F 2018 *Front Mater.* **5** 16.
[30] Amendola A and Fraternali F 2018 *Compos Struct.* **191** 145-53.

IOP Conf. Series: Materials Science and Engineering **999** (2020) 012010 doi:10.1088/1757-899X/999/1/012010

Mechanical characterization of FDM filaments with PVDF matrix reinforced with Graphene and Barium Titanate

I Farina[1], R Sharma[2], R Singh[3], A Batish[2], N Singh[4], F Fraternali[4] and F Colangelo[1]

1Dept. of Engineering, University of Naples (Italy)
2Dept. of Mechanical Engineering, Thapar Institute of Engineering and Technology, Patiala (India)
[3]Dept. of Mechanical Engineering, National Institute of Technical Teachers Training and Research, Chandigarh (India)
[4]Dept. of Civil Engineering, University of Salerno (Italy)
Email: ilenia.farina@uniparthenope.it; sharmaravi2190@gmail.com; rupindersingh78@yahoo.com; abatish@thapar.edu; f.fraternali@unisa.it; francesco.colangelo@uniparthenope.it

Abstract: In past one decade number of studies has been reported on optimization of process parameters of fused deposition modelling (FDM) for in-house developed thermoplastic composite based feed stock filaments. This paper investigates smart polymer-based composites prepared with hybrid feed stock filament (comprising of polyvinylidene fluoride (PVDF) reinforced with graphene (Gr) and barium titanate (BTO) powder). This work started with the Functional prototypes were 3D printed for tensile and flexural characterization using inhouse developed filament (PVDF (78%)+Gr (2%) with BTO (20%)) at optimized settings of FDM. The printed specimens were subjected to destructive testing for mechanical properties (to analyze the process capability indices, Cp and Cpk). For morphological properties, scanning electron microscopy (SEM) images and 3D rendered images of the fractured surfaces of tensile and flexural specimen were used. It has been revealed from the SEM and 3D rendered images that the optimized settings of 3D printing process parameters resulted into uniform morphological features (based upon surface roughness (Ra) and amplitude distribution function (ADF), peak count (PC) and bearing ratio (BR) curve).

1. Introduction

The FDM is an additive manufacturing (AM) process in which 3D physical models are built up by joining the materials in the form of layers as per computerized created 3D solid model and it has ability to produce minimal wastage with acceptable level of accuracy [1]. This creates the possibility to fabricate parts having complex geometry that are difficult to produce via conventional material removal processes [2]. AM does not depend upon the fixtures, coolants, cutting tools, and other conventional resources for the production of parts [3]. In FDM, a hot nozzle-based extruder is used for melting and depositing of semi molten material to build a 3D physical part. FDM generally taking a thermoplastic or polymer matrix composite (PMC) filament as a printing material [4]. The open source FDM printers provide the simple operating conditions and good control over the input variables such as infill speed, infill angle, infill density, raster width, printing temperature, type of fill etc. [5-8]. By varying these printing parameters, it is possible to produce customized parts on demand and allows to optimize the product design [9]. Research studies highlighted that the properties of fabricated parts are function of process variables and can be efficiently improved with appropriate adjustment. For functional prototypes the mechanical properties are very important, also it is essential to study the effect of process variables on the mechanical properties of the parts which can be efficiently improved by process optimization [10-13]. Due to wide acceptability of this process and ease of availability of 3D printers, researchers started work to develop new materials for FDM process. So far various polymer matrix composites-based feed stock filaments have been developed by reinforcement of different metal powders or ceramics to improve the inherent properties of the polymer matrices [14-15]. Some of the majorly used polymers as printing material on FDM are acrylonitrile butadiene styrene (ABS), poly lactic acid (PLA), polyamide (PA6), polyvinyl chloride (PVC) or the composites based upon these polymers.

In the family of electroactive polymers PVDF is mostly used in piezoelectric sensor applications [16-18]. Similarly, the addition of carbon fibres increased the mechanical properties of nylon-based polymer composite matrix. Some researchers have reported the addition of Gr in various polymer matrixes for the development feedstock filament of FDM. It has been observed that the addition of Gr improved the mechanical, electrical and thermal characteristic of base polymeric materials [18]. Some researchers explored the effect of FDM input parameters on the mechanical properties of the 3D printed parts [19]. This study is the extension of previously conducted research work [17,18] in which electroactive polymer composite matrix of PVDF was 3D printed and process parameters of FDM were optimised, but the repeatability/reproducibility and statistical control of process was not addressed. This research work presents the process capability analysis of FDM processing parameters for mechanical properties such as peak strength, break strength and modulus of toughness of electroactive polymer composites-based 3D printed tensile and flexural specimens. As this study reports an extension of previously reported work [18,19] in which a composite of fix proportion of BTO 20% (wt.%) in PVDF (78%) +Gr (2%) was prepared for 3D printing of functional graded prototypes. therefore, no changes have been made in composition of PVDF based composites for 3D printing and similar optimized conditions were used for FDM processing.

2. Experimentation

2.1. *Materials and methods*

In this study, Gr and BTO were reinforced in PVDF matrix for the fabrication of feed stock filament. A twin screw extruder (TSE) (Model: Mini CTW, HAAKE, Germany) was used for compounding of materials and extrusion of feed stock filament. In the previously conducted research, it has been reported that BT (20%) in PVDF (78%)+Gr (2%) has shown best mechanical properties when extruded at optimized settings of TSE (extrusion temperature 200°C and screw speed 50 rpm) [19]. Therefore in this research work the same proportion of the composition has been taken for the preparation of feed stock filament for 3D printing of the standard specimens. The key point while extruding the feedstock filament is to keep the diameter of the wire within the range of 1.75±0.05 mm (due to the restricted nozzle size available in the existing FDM setup). Finally, the extruded feedstock filament was used to run on an open source FDM printer (Make: Divide by Zero, India) for 3D printing of tensile specimens (as per ASTM D-638) and flexural specimens (as per ASTM D-790) at standard optimized processing conditions (infill speed: 50 mm/s, infill angle: 45° and infill density: 100 %).

Ten standard tensile and flexural specimens were 3D printed to report the process capability analysis of 3D printing process parameters. The 3D printed functional prototypes were subjected to universal testing machine (UTM) for destructive testing. The output results of mechanical testing of both type of specimens in the form of peak strength (PS), break strength (BS) and fracture toughness (FT) were noted to perform process capability analysis. Further SEM images were observed for morphological characterization of tensile and flexural specimens tested on UTM. The microscopic images were further used for 3D rendering to generate Ra profile, ADF, PC and BR curve.

3. Results and discussions

3.1. *Mechanical testing*

IOP Conf. Series: Materials Science and Engineering **999** (2020) 012010 doi:10.1088/1757-899X/999/1/012010

After successfully completion of 3D printing of tensile and flexural specimens on the open source FDM setup at standard optimized conditions, destructive testing was performed. The fabricated samples were mechanically tested on UTM (Make: Shanta engineering, Pune, India; Capacity: 5000N).

The output UTM results for tensile and flexural properties in terms of PS, BS and FT are shown in Table 1. The resulted output values of tensile and flexural properties were further used for statistical process capability analysis.

Table 1.UTM results for tensile (left) and flexural (right) specimens

S.No.	PS (MPa)	BS (MPa)	FT (MPa)
1	42.95	38.70	3.70
2	43.10	39.02	3.54
3	42.68	37.93	4.27
4	41.98	37.04	4.38
5	42.38	37.82	4.54
6	43.19	38.67	3.79
7	43.40	38.92	3.68
8	42.54	37.84	4.14
9	43.14	40.01	4.72
10	42.25	37.24	3.28

3.2. Morphological analysis

The fractured surfaces of tensile and flexural specimen were subjected to SEM for morphological properties. Further, the SEM images of fractured surface were 3D rendered using image processing software package tool. Figure 1(a) and 2(a) shows the SEM images of tensile and flexural specimen. Whereas, the figure 1(b) and 2(b) shows the rendered 3D images of tensile and flexural specimen. Based upon 3D rendered images, plots for Ra profile, ADF, PC and BR curve of tensile and flexural specimens were generated (Figure 1 and 2). As plots drawn from 3D rendered images for tensile and flexural specimens represented the similar variations in Ra profile, ADF, PC and BR curve of tensile and flexural specimen, which justifies the selection of optimized settings of parameters and controlled statistical nature of process.

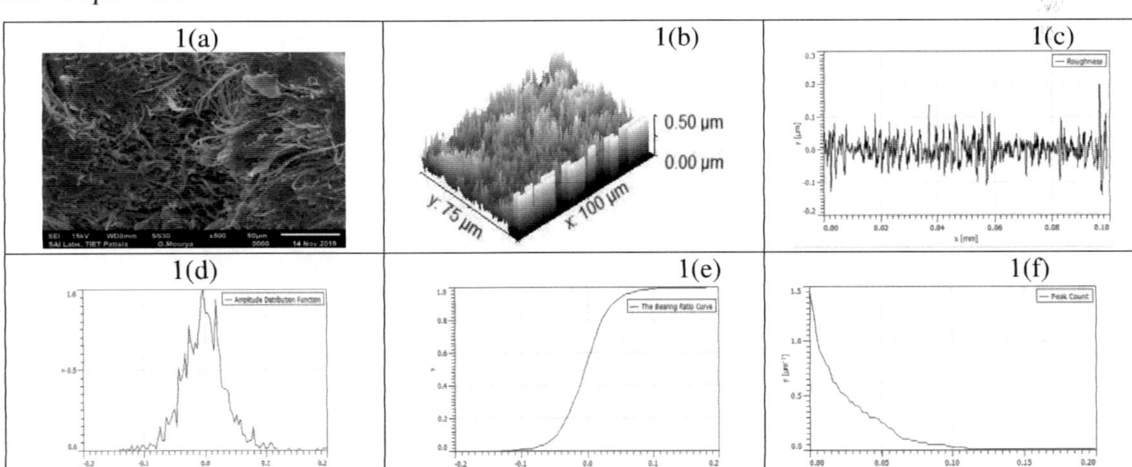

Figure1. (a)SEM image of fractured surface, (b) 3D rendered image, (c) Ra profile, (d) AFD, (e) BR curve, (f) PC of tensile specimens.

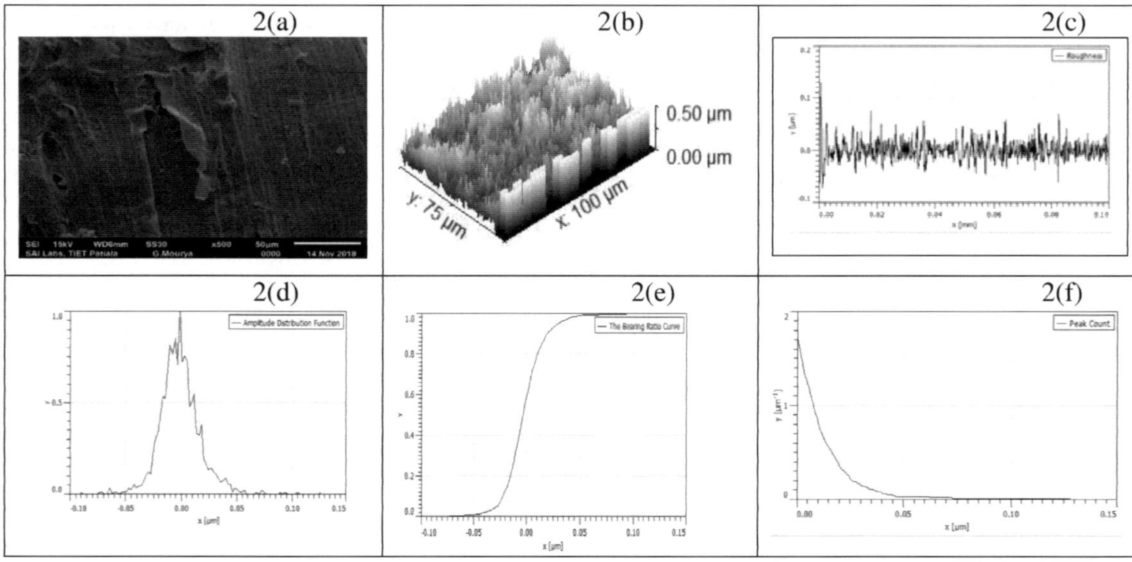

Figure2. (a)SEM image of fractured surface, (b) 3D rendered image, (c) Ra profile, (d) AFD, (e) BR curve, (f) PC of flexural specimens.

The Ra value for tensile specimen was 27.79 nm at cut-off length of 0.05 whereas Ra value for flexural specimen was observed as 13.63 nm on the same cut-off length. Thus, there is not much variation among the Ra values of both specimens. Similarly, the curves of ADF (Figure 1(d) and 2(d)), BR (figure: (1(e) and 2(e)) (for length of the bearing surface at specified depth in the evaluation area) and PC (figure: 1(f) and 2(f) for both the cases follows the same trend.

4. Conclusions

Following are the conclusions from this study:

- From process capability indices the observed values of Cp and Cpk were greater than 1 for both cases clearly represented that FDM printing of PVDF+Gr+BTO based composite at optimized parametric settings resulted into statistically controlled process.
- It has been ascertained from the SEM images and 3D rendered images that the optimized 3D printing parameters resulted into uniform morphological features (as per Ra, ADF, PC and BR curve) for both tensile and flexural samples.

In future studies, we will investigate the design of novel sustainable composites [20-24] using waste and alternative materials as precursors [25-29] and we will assess their environmental impact [30-31]. Additional future research lines will investigate the use of eco-sustainable materials for the design of next-generation meta-materials and structures, whose properties mainly derive from the geometry of the microstructure [32-35].

References

[1] Onwubolu GC and Rayegani F 2014 *Int. J. of manufacturing Eng.*
[2] Mohan N, Senthil P, Vinodh S and Jayanth N 2017 *Virtual and Phys Prot* **12(1)** 47-59.
[3] Huang SH, Liu P, Mokasdar A and Hou L 2013 *The Int. J of Adv Manufacturing Tech* **67(5-8)** 1191- 03.
[4] Wang X, Jiang M, Zhou Z, Gou J and Hui D 2017 *Comp. Part B: Eng* **110** 442-58.
[5] Asadollahi-Yazdi E, Gardan J, and Lafon P, 2018 *The Int J of Adv Manufacturing Tech* **96 (9-12)** 4145-64.
[6] Singh S and Singh R 2020 *Rapid Prototyping Journal.*
[7] Singh R, Sharma R, and Ranjan N 2017 doi:10.1016/B978-0-12-803581-8.10167-5.
[8] Thompson MK, Moroni G, Vaneker T, Fadel G, Campbell RI, Gibson I, Bernard A, Schulz J, Graf P Ahuja B and Martina F 2016 *CIRP annals* **65(2)** 737-60.
[9] Mangat AS, Singh S, Gupta Mand Sharma R, 2018 *Rapid Prototyping Journal.*

[10] Sood AK, Ohdar RK and Mahapatra SS 2010 *Materials & Design* **31(1)** 287-95.
[11] Mohamed OA, Masood SH and Bhowmik JL 2015 *Adv.in Manufacturing*, **3(1)** 42-53.
[12] Vijayaraghavan V, GargA, Lam JSL, Panda B and Mahapatra SS 2015 *The Int J of Adv Manufacturing Tech* **78(5-8)** 781-93.
[13] Singh R, Sharma R and Davim JP 2018 *Int J of Prod Res* **56 (24)** 7330-40.
[14] Roberson DA, Espalin D and Wicker RB 2013 *Virtual and Physical Prototyping* **8(3)** 201-12.
[15] Sharma R, Singh R and Batish A 2020 *Materials Today: Proceedings*.
[16] Singh R, Sandhu GS, Penna R and Farina I 2017 *Materials* **10(8)** p.881.
[17] Sharma R, Singh R. and Batish A 2019 *J. of Thermoplastic Comp Materials* p.0892705719865004.
[18] Sharma R, Singh R and Batish A 2020 *Journal of Thermoplastic Composite Materials* p.0892705720907677.
[19] Fraternali F, Farina I and Carpentieri G 2014 *Mech Res Commun* **56** 18-25.
[20] Fabbrocino F, Farina I and Modano M 2017 *Compos Part B-Eng* **115** 376-83.
[21] Fabbrocino F, Farina I, Amendola A, Feo L and Fraternali F 2016 ECCOMAS Congress 2016 – Proceedings of the 7th European Congress on Computational Methods in Applied Sciences and Engineering, 1, 1893-1908.
[22] Kumar R, Singh R, Hui D, Feo L and Fraternali F 2018 *Compos Part B-Eng.* **134**, 193-06.
[23] Singh, R, Kumar R, Farina I, Colangelo F, Feo L and Fraternali F 2019 *Polym J* **11**(1), 62.
[24] Roviello G, Menna C, Tarallo O, Ricciotti L, Ferone C, Colangelo F and Cioffi R 2015 *Mater Des* **87**, 82-94.
[25] Singh R, Singh H, Farina I, Colangelo F and Fraternali F 2019 *Compos Part B-Eng* **156**, 259-65. Colangelo F, Cioffi R, Roviello G, Capasso I, Caputo D, Aprea P and Ferone C 2017 Compos Part B-Eng **129**, 11-17.
[26] Roviello G, Ricciotti L, Tarallo O, Ferone C, Colangelo F, Roviello V and Cioffi R 2016 *Materials* **9** (6), 461.
[27] Colangelo F, Petrillo A, Cioffi R, Borrelli C and Forcina A 2018 *Sci Total Environ*, **615** 1506-17.
[28] Colangelo F, Forcina A, Farina I and Petrillo A 2018 *Buildings* **8**(5) 70.
[29] Fraternali F and Feo L 2000 *Compos Part B-Eng.* **31** 141-58.
[30] Fortunato A, Fraternali F and Angelillo A 2014 *Ing Sismica-Ital.* **31**(1) 41-51
[31] Modano M, Fabbrocino, F, Gesualdo A, Matrone G, Farina I and Fraternali F 2015 COMPDYN 2015—5th ECCOMAS Thematic Conference on Computational Methods in Structural Dynamics and Earthquake Engineering 209-17.
[32] Fraternali F, Spadea S and Ascione L 2013 *Compos Struct.* **100** 280-89.
[33] Fabbrocino F, Amendola A, Benzoni G and Fraternali, F 2016 *Ing Sismica-Ital* **1-2**, 62-71.
[34] Mascolo I, Amendola A, Zuccaro G, Feo L and Fraternali F 2018 *Front Mater* **5** 16.
[35] Amendola A and Fraternali F 2018 *Compos Struct.* **191** 145-53.

IOP Conf. Series: Materials Science and Engineering **999** (2020) 012011 doi:10.1088/1757-899X/999/1/012011

Simulation and Experiment study on lubrication and fatigue properties of white alloy journal bearings

Ning Ding[1], Hulin Li[2,*], Zhongwei Yin[2], Ning Zhong[3], Yonghong Chen[3]

[1] School of Mechanical Engineering, Shanghai Jiao Tong University, 800 Dongchuan RD, Minhang District, Shanghai 200240, PR China
[2] School of Design, Shanghai Jiao Tong University, 800 Dongchuan RD, Minhang District, Shanghai 200240, PR China
[3]Chongqing Yuejin Machinery Co., Ltd., No.1 Huagong RD, Yongchuan District, Chongqing 402169, PR China
*Corresponding author: E-mail address: lhl666@sjtu.edu.cn (Hulin Li)

Abstract. In this paper, the elastohydrodynamic (EHD) simulation by AVL EXCITE is performed to analyze the lubrication performance including oil film thickness, oil film pressure, asperity contact pressure, and friction power losses. The journal bearing fatigue experiments were conducted on the fatigue test rig (Glacier Vandervel, made in the UK). The shell temperature was measured in real-time and the fatigue limit was finally determined. The simulation results combined with experimental results were discussed to comprehensively explain the failure mechanisms. It can be found that as the lubrication condition degrades, the asperity contact aggravates, and the shell temperature is increased sharply. This causes bearing performance degradation, poor bearing load carrying capacity, and severe surface damage, which finally lead to bearing failure.

Keywords: EHD; AVL EXCITE; Journal bearing; Lubrication performance; Fault mechanism

1. Introduction

Journal bearings are widely used due to their stability, reliability, and noiselessness, which are one of the critical elements in rotary machines. During operation, long-term high cyclic load, poor lubrication performance, and high shell temperature may become the main causes of bearing faults. It is necessary to combine the simulation and the experiment to comprehensively observe the lubrication condition and operation state to analyze the fault mechanisms to avoid as much as possible severe disasters.

Generally, the shaft and the journal bearing are both elastic bodies, and the surface of the shaft and the bearing will produce elastic deformation when subjected to force, which will affect the lubrication characteristics of the journal bearing. Therefore, the elastohydrodynamic (EHD) simulation analysis has been developed to evaluate the lubrication performance by many related studies. D.E. Sander et al. [1] and Florian König et al. [2] used the EHD simulation to evaluate the lubrication condition of journal bearing considering the lubricating oil properties, the elastic deformation, and surface topography parameters of the bearing and the shaft. C. Priestner et al. [3] and D.E. Sander et al. [4] evaluated the lubrication performance and wear behavior under different cyclic loads by EHD simulation. H. Allmaier et al. [5] used the EHD analysis to investigate the influence of the shell temperature on the lubrication condition. Besides, the authors [6] also applied the EHD analysis to study the relationship between the lubrication condition and friction behavior.

However, although most studies are widely proposed for the journal bearing lubrication performance evaluation, few researchers have verified the simulation results through experiments to comprehensively discuss the internal damage mechanism and failure causes. In this paper, the EHD analysis is firstly achieved to evaluate the journal bearing lubrication condition. Then the fatigue experiment is conducted to observe the surface damage phenomena. Based on the simulation and experiment results, the damage mechanisms of journal bearings are discussed to provide technical support for the actual application of failure analysis.

2. Theory

2.1. Averaged Reynolds equation

The full oil film interaction is solved by the averaged Reynolds equation, which considers the influence of roughness amplitudes. The specific formula is as follows [7]:

$$\frac{\partial}{\partial x}\left(-\frac{h_T^3}{12\eta}\frac{\partial P}{\partial x}+\frac{u_1+u_2}{2}h_T\right)+\frac{\partial}{\partial y}\left(-\frac{h_T^3}{12\eta}\frac{\partial P}{\partial y}\right)+\frac{\partial h_T}{\partial t}=0 \tag{1}$$

where h_T denotes the real clearance considering roughness heights of journal and bearing. P is the hydrodynamic pressure and x, y represents the circumferential and axial direction respectively. H indicates the oil viscosity.

2.2. Lubricating oil properties

The Vogel equation [8] is applied to describe the relationship between the oil viscosity and oil temperature in the simulation model:

$$\eta(T)=Ae^{\frac{B}{T+C}} \tag{2}$$

where A=0.198 mPa·s, B=736.69 °C, C=88.9 °C for oil Shell Rimula R3 Multi 10W-30 in this paper. T is the oil temperature.

2.3. Asperity contact interaction

Asperity contact interaction exists when asperity summits of two nominal flats rub each other. The Greenwood and Tripp model is used to evaluate the behavior of asperity contact. The asperity contact pressure is calculated as [9]:

$$P_a=KE^*F_{5/2}(H_s) \tag{3}$$

where the elastic factor K is set to 0.003. $F_{5/2}(H_s)$ stands for the form function and E^* is the composite elastic modulus, as listed in Eq. (4) and (5):

$$F_{5/2}(H_s)=\begin{cases}4.4086\times10^{-5}(4-H_s)^{6.804} & H_s<4\\ 0 & H_s\geq 4\end{cases} \tag{4}$$

$$E^*=\frac{1}{\left(\frac{1-v_1^2}{E_1}+\frac{1-v_2^2}{E_2}\right)} \tag{5}$$

where H_s is the non-dimensional summit clearance, v_1 and v_2 are Poisson ratios of bearing material and shaft material respectively, E_1 and E_2 are Young moduli of bearing material and shaft material respectively.

2.4. Lubrication conditions

Generally, the film thickness ratio is designed as a criterion for the determination of the lubrication condition. It is calculated as follows [10]:

$$\lambda=\frac{h_{\min}}{\sqrt{\sigma_1^2+\sigma_2^2}} \tag{6}$$

IOP Conf. Series: Materials Science and Engineering **999** (2020) 012011 doi:10.1088/1757-899X/999/1/012011

where h_{min} represents the minimum oil film thickness. σ_1 and σ_2 are surface roughness of bearing and journal respectively. According to the range of the film thickness ratio, journal bearing lubrication regimes are divided into three kinds: boundary lubrication ($\lambda \leq 1$), mixed lubrication ($1 < \lambda < 3$), hydrodynamic lubrication ($\lambda \geq 3$).

3. Simulation

Generally, the bearing and the shaft are fully separated by the oil film at the hydrodynamic lubrication. Due to the increase of the external load or the oil starvation, the oil film may be broken eventually resulting in the direct contact between the bearing and the shaft. The asperity contact pressure will play a dominant role in supporting the bearing operation. The lubrication condition gradually transforms from hydrodynamic lubrication to mixed lubrication or boundary lubrication. Based on the principle, in AVL EXCITE, the interaction in the simulation model including total normal pressure, tangential stress, and energy dissipation are divided into two types: full oil film interaction and asperity contact interaction. The simulation model comprises the test connecting rod and the test bearing. The model information about the geometry, mass, stiffness, and degree of freedom is condensed to the FE condensation model. The simulation model specimen is presented in Figure 1. The simulation parameters are listed in Table 1. The test loads for the maximum external load of 20MPa, 27MPa, 34MPa are applied in the simulation model, as shown in Figure 2(a). The simulation results are listed in Table 2.

Table 1. Simulation parameters.

Parameter	Value
Bearing width (mm)	29.5
Bearing wall thickness (mm)	1.825
Bearing clearance (mm)	0.04
Bearing material	ZSnSb11Cu6
Elastic modulus of bearing material (GPa)	47.93
Poisson ratio of bearing material	0.2847
Bearing surface roughness (μm)	0.4
Shaft diameter (mm)	52.7
Shaft material	42CrMo
Elastic modulus of shaft material (GPa)	212
Poisson ratio of shaft material	0.28
Shaft surface roughness (μm)	0.2
Shaft Speed (rpm)	3000
Maximum external load (MPa)	20/27/34
Lubricant grade	Shell Rimula R3 Multi 10W-30
Lubricating Oil supply pressure (MPa)	0.6
Lubricating Oil supply temperature (°C)	70

Table 2. Simulation results.

parameter	Maximum external load		
	20MPa	27MPa	34MPa
MOFT (μm)	1.87	1.24	0.94
POFP (MPa)	58.2	78	93.8
PASP (MPa)	0	54.7	103
Total friction power losses (W)	170	228	403
Film thickness ratio	4.156	2.756	2.089
Lubrication condition	Hydrodynamic lubrication	Mixed lubrication	Mixed lubrication

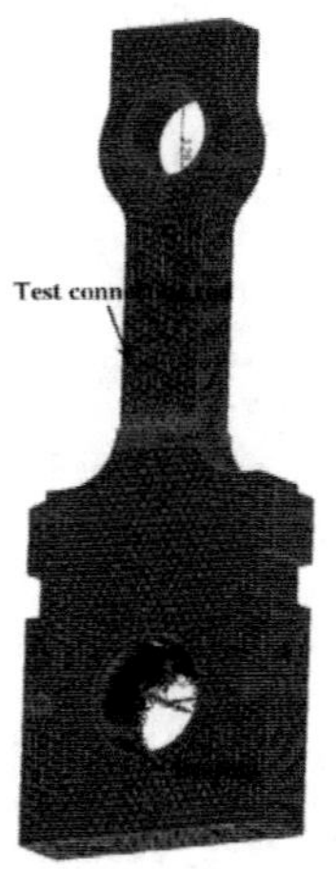

Figure 1. The simulation model.

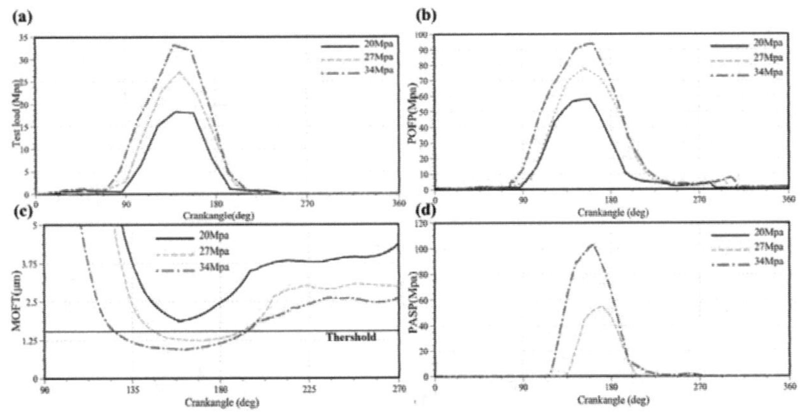

Figure 2. The simulation results: (a): the test load; (b): the peak oil film pressure (POFP); (c): the minimum oil film thickness (MOFT); (d): the peak asperity contact pressure (PASP).

Figure 2(b) shows the plot of the peak oil film pressure (POFP). With the increase of the maximum external load, the POFP improves from 58.2MPa to 93.8MPa to enhance the carrying load capacity. Simultaneously, the minimum oil film thickness (MOFT) declines correspondingly in Figure 2(c). According to Eq. (7), the MOFT of 1.4μm is the basis to judge whether the asperity contact occurs. For 20Mpa, the MOFT of 1.82μm is higher than the threshold, thus there is no asperity contact pressure. The test bearing runs in the hydrodynamic lubrication and has good performance. The total pressure is entirely provided by the hydrodynamic pressure. The film thickness ratio is calculated as 4.156, which satisfies the demand of the hydrodynamic lubrication condition. Nevertheless, the MOFT is 1.24μm and 0.94μm respectively for 27MPa and 34MPa, which are lower than the threshold, The oil film is broken and asperity contact between the shaft and the bearing occurs. The bearing operating condition transforms from hydrodynamic lubrication to mixed lubrication. The total pressure consists of the hydrodynamic pressure and the asperity contact pressure. It can be verified that the film thickness ratio is 2.756 and 2.089 respectively for 27MPa and 34MPa. The bearing lubrication condition belongs to the mixed lubrication.

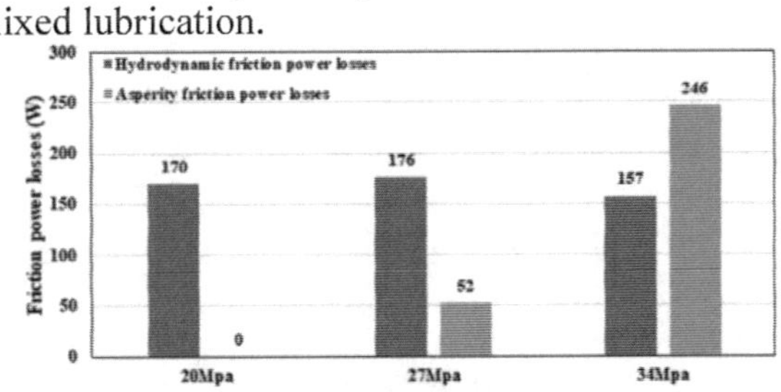

Figure 3. Comparison of the friction power losses for three load cases.

Figure 3 demonstrates the friction power losses condition under three external loads. Total friction power losses are classified into two aspects: hydrodynamic power losses generated from hydrodynamic pressure and power losses because of asperity contact. For 20MPa, no asperity contact pressure exists, hence asperity friction power loss is zero and total friction power losses only include hydrodynamic power losses, with a value of 170W. For 27MPa, due to the occurrence of asperity contact pressure, 52W asperity friction power losses generate, which account for 22.8% of total power losses. With the asperity contact pressure increasing, power losses generated from asperity contact increase to 246W, which are 61% of total power losses for 34MPa. Thus it can be found that the appearance of the asperity contact improves the asperity contact power loss ratio. Friction power

IOP Conf. Series: Materials Science and Engineering **999** (2020) 012011 doi:10.1088/1757-899X/999/1/012011

losses usually dissipate in the term of heat, which causes the local high temperature on the rubbing area of the bearing surface and the deterioration of material performance.

The contour of the PASP distribution at different crank angles for three load cases is investigated in Figure 4. Three crank angle 120°, 160°, 200° are chosen to describe the variation on the asperity contact distribution. For 27MPa, the slight asperity contact remains at one edge at crank angle 120°. At crank angle 160°, the asperity contact exists at both edges. At crank angle 200°, there appears the mild asperity contact at one edge and in the central region of the bearing. For 34MPa, the PASP is increased. Besides, the asperity contact area spreads from one side to both sides at crank angle 120°, from the bearing edges to the central area at crank angle 200°. Generally, these areas may inevitably suffer to wear. Serious wear will endanger directly the long term normal operation of bearings.

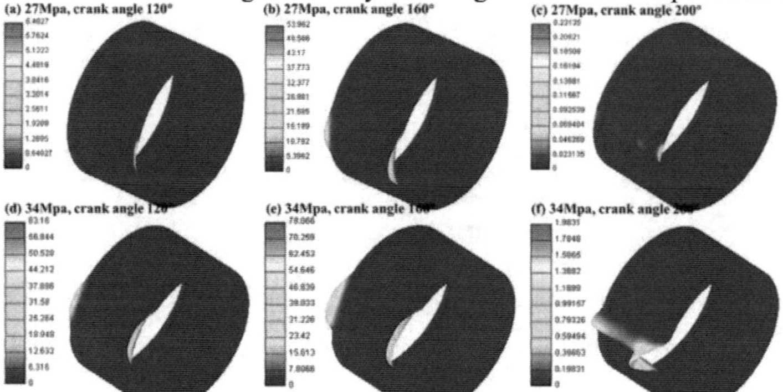

Figure 4. The contour of the PASP distribution at different crank angles for three load cases.

4. Experiment

The bearing specimens are obtained by centrifugal casting, which were made of the bimetal structure containing a ZSnSb11Cu6 alloy substrate layer with a thickness of 0.6mm and the 10#steel back. The chemical compositions of bearing specimens are listed in Table 3. To verify the simulation results, the fatigue tests were conducted on the journal bearing fatigue test-rig at Glacier Vandervall, as sketched in Figure 5. The load-increasing method is applied to measure the bearing fatigue limit. Firstly, the test bearing is run for 30 minutes without load. Secondly, the initial load is set to 20MPa, the test-rig continues to run for 20h (equivalent to 3.6×10^6 load cycles), if no damage on the bearing surface is observed, the initial load will be increased by 7MPa. The test is repeated until the occurrence of bearing surface damage or the test-rig load limit. The bearing fatigue limit is determined as the previous load or the test-rig load limit. During the test, the shell temperature is measured in real-time by the temperature sensor.

Table 3. The chemical compositions of bearing specimens.

Material							
ZSnSb11Cu6	Chemical composition	Sn	Pb	Sb	As	Cu	Fe
	Value	Bal.	0.35	10-12	0.1	5.5-6.5	0.1
10#Steel	Chemical composition	C	Si	Mn	Cr	Ni	Cu
	Value	0.07-0.13	0.17-0.37	0.35-0.65	≤0.15	≤0.30	≤0.25

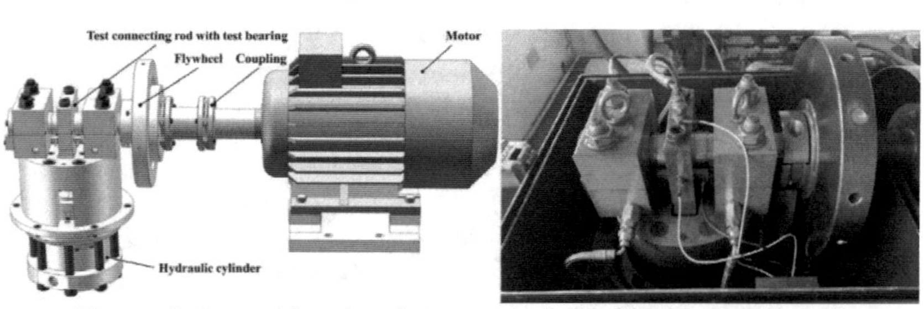

Figure 5. Journal bearing fatigue test-rig at Glacier Vandervall.

IOP Conf. Series: Materials Science and Engineering **999** (2020) 012011 doi:10.1088/1757-899X/999/1/012011

5. Discussion

5.1. Surface damage condition

Figure 6 shows the test bearing surface damage conditions for three load cases. For 20MPa, no obvious damage is observed on the surface of the testing bearing, which represents that the testing bearing may work normally during this stage. Combined with the simulation results, the bearing runs in the hydrodynamic lubrication, good lubrication performance does not cause bearing fatigue failure. For 27MPa, it can be seen clearly that the fatigue wear exists at one edge of the bearing. According to the above simulation results, the bearing runs in the mixed lubrication, the presence of the asperity contact pressure marks that the rubbing occurs between the bearing and the shaft. Because the rubbing area locates at one edge, so slight damage is eventually formed. While for 34MPa, the bearing surface is seriously damaged. A lot of peelings and severe material adhesion appear on the surface of the bearing. Based on the simulation results, it can be found that the lubrication condition of the bearing is the mixed lubrication and the asperity contact pressure increases sharply in the rubbing region, which leads to the increase of asperity friction power losses. This causes high shell temperature in the rubbing region, which makes the material performance degenerate due to the low melting point of the ZSnSb11Cu6 alloy. Consequently, the material adhesion appears on the surface of the bearing. In addition, the comprehensive influence of cycle load and asperity contact pressure promotes the initiation and propagation of fatigue cracks and finally the formation of peelings. Based on the above analysis, the fatigue limit is determined as 27MPa.

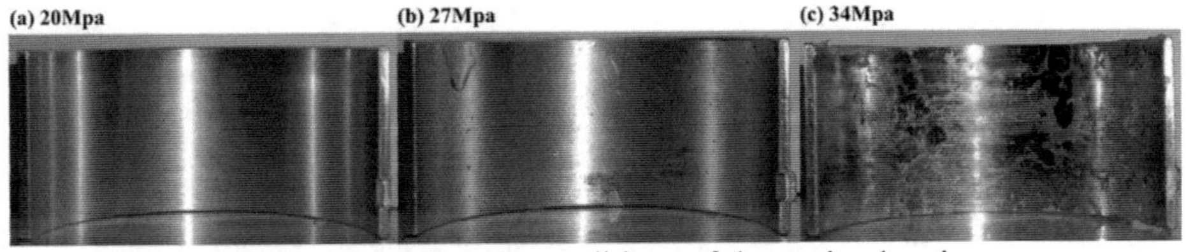

Figure 6. The surface damage conditions of the testing bearing.

5.2. Shell temperature

The shell temperature is recorded after the testing bearing runs for 2.5 hours in Figure 7. The curves reflect that the temperature gradually rises with time and then stabilizes. And the temperature changes within 80°C~90°C for 20MPa and 27MPa. The friction power loss dissipation in the form of heat is a consequence of the increase in shell temperature. Based on simulation results, no asperity contact power loss is generated for 20MPa, while only 52W asperity contact power losses are produced for 27MPa. Consequently, the shell temperature of 27MPa is higher than that of 20MPa. No severe adhesive damage appears on the bearing surface. However, the temperature increases sharply to 103°C for 34MPa. The high shell temperature promotes the melting of ZSnSb11Cu6 alloy and the degradation of bearing performance, which is an important cause of bearing failure.

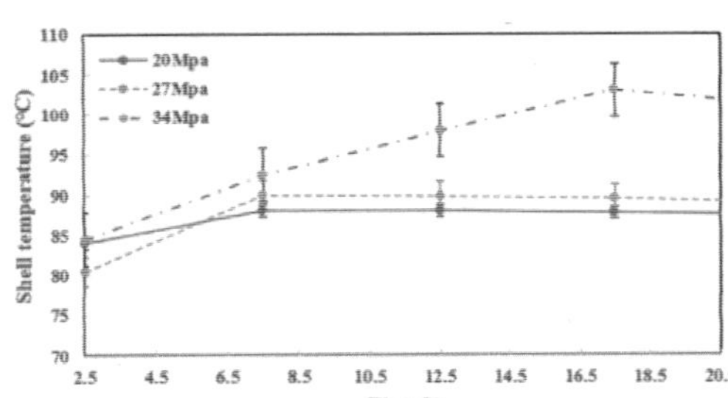

Figure 7. The shell temperature during the fatigue test.

6. Conclusion

In this paper, the lubrication performance of the journal bearing cover with a ZSnSb11Cu6 alloy is studied by AVL EXCITE. Combined with the simulation results, the fatigue tests are conducted to analyze the damage mechanisms of journal bearing. The main conclusions are as follows:

(1) For 20MPa, no damage is observed on the surface of the bearing. The simulation result indicates that the bearing runs in hydrodynamic lubrication. The experiment results are consistent with the simulation results.

(2) For 27MPa, the fatigue damage occurs on one side of the bearing, which means that the bearing lubrication regime transforms into the mixed lubrication and the asperity contact appears. This conclusion is also consistent with the simulation analysis.

(3) For 34MPa, the bearing still runs in mixed lubrication. Under the comprehensive effect of cycle load and asperity contact pressure, the material adhesion and fatigue damage exist on the surface of the bearing, eventually leading to bearing failure.

(4) As the asperity contact aggravates, the shell temperature rise caused by friction power losses increases significantly, which causes bearing performance degradation, poor bearing load carrying capacity, and bearing failure.

Acknowledgement

The authors acknowledge the financial support provided by Marine Low Speed Engine Project-Phase I (Grant No. CDGC01-KT11).

References

[1] D.E. Sander, H. Allmaier, H.H. Priebsch, F.M. Reich, M. Witt, T. Füllenbach, A. Skiadas, L. Brouwer, H. Schwarze, Impact of high pressure and shear thinning on journal bearing friction, Tribology International, 81 (2015) 29-37.

[2] F. König, A. Ouald Chaib, G. Jacobs, C. Sous, A multiscale-approach for wear prediction in journal bearing systems – from wearing-in towards steady-state wear, Wear, 426-427 (2019) 1203-1211.

[3] C. Priestner, H. Allmaier, H.H. Priebsch, C. Forstner, Refined simulation of friction power loss in crank shaft slider bearings considering wear in the mixed lubrication regime, Tribology International, 46 (2012) 200-207.

[4] D.E. Sander, H. Allmaier, H.H. Priebsch, F.M. Reich, M. Witt, A. Skiadas, O. Knaus, Edge loading and running-in wear in dynamically loaded journal bearings, Tribology International, 92 (2015) 395-403.

[5] H. Allmaier, C. Priestner, F.M.R. a, H.H. Priebsch, Novotny-Farkas, Predicting friction reliably and accurately in journal bearings—extending the EHD simulation model to TEHD, Tribology International, 58 (2013) 20-28.

[6] H. Allmaier, C. Priestner, C. Six, H.H. Priebsch, C. Forstner, F. Novotny-Farkas, Predicting friction reliably and accurately in journal bearings—A systematic validation of simulation results with experimental measurements, Tribology International, 44 (2011) 1151-1160.

[7] P. N, C. H, An average flow model for determining effects of threedimensional roughness on partial hydrodynamic lubrication, ASME, Trans J Lubric Technol 100 (1978) 12-17.

[8] V. H, The law of the relation between the viscosity of liquids and the temperature, Physica Z 22 (1921) 645-646.

[9] G. J, W. J, Contact of nominally flat surfaces., Proceedings of the Royal Society of London. Series A, Mathematical and Physical Sciences, 295 (1966) 300-319.

[10] B. J, Hamrock, Fundamentals of Fluid Film Lubrication, NASA Reference Publication, (1991).

Automatic Surface Defect Inspection System Using Convolutional Neural Networks

X Zhu[1,2*], S K Kundu[1,3*], and N K Bangalore Ramaiah[1,4]

[1]Hitachi America, Ltd. Research & Development Division, Farmington Hills, MI, 48335 USA
[2]xiaoliang.zhu@hal.hitachi.com, [3]subrata.kundu@hal.hitachi.com,
[4]naveenkumar.bangaloreramaiah@hal.hitachi.com
*Corresponding authors: xiaoliang.zhu@hal.hitachi.com, subrata.kundu@hal.hitachi.com

Abstract. Surface quality of the piston rod is very important for the durability of the shock absorbers. However, currently in production, it is very difficult to use the available systems to inspect the whole rod surface online, in real time (~4s), with a high sensitivity and accuracy. To overcome this, in this paper, an online automatic rod inspection system has been developed, which allows to inspect the whole rod surface in 4s, with a sizing accuracy around 75%, and detection accuracy >90%. An integrated software is used for rod rotation control, image capturing, image processing, and decision making. Convolutional Neural Network is used to processing the 360° surface image with a high accuracy, eliminating errors caused by environmental lighting. A new method based on aspect ratio and size information is used for defects classification. Experimental results show that the system is capable of detecting defects as small as 25μm and differentiating nodule, dent, and scratch with a processing time around 4s per rod.

Keywords: Surface quality; machine vision; CNN; dent; nodule; scratch

1. Introduction

Automotive suspension system, which can be classified as passive, semi-active, and active suspension, plays a more and more important role in driving stability and comfort by reducing or eliminating the vibrations transmitted to vehicle body [1]. As one of the key moving component of the shock absorbers in the suspension system, the piston rod is forced moving reciprocally by the vibrations or bumps from the road, moving hydraulic oil between different chambers in the absorber and converting the vibration energy into heat which is then dissipated. During the piston rod's reciprocal movement, if there are any defects on the rods surface, namely, nodule, dents, and scratches, the hydraulic oil seal will be quickly damaged, causing oil leakage, noise, and eventually a failure of the suspension system. Thus, it is critical to inspect the rod surface quality before assembling.

The traditional way to inspect the rod surface is performed manually by either visual inspection (eyeballing the rod surfaces) or scratching rod surfaces with a knife. For the latter, by feeling the resistances caused by defects, defective rods can thus be screened. Although simple, those methods not only highly rely on workers experience which is subjective and labor-intensive, but also time consuming and raise risk of surface damage due to the additional manual handling. In an attempt to solve this problem, some non-destructive online sensors have been developed. Ismagilov et al. have developed a method for detecting surface defects in micron-thickness metallic films by scanning the object surface with a focused laser beam. A piezoelectric sensor is then used to detect the generated acoustic waves (Raleigh wave) on the object surface [2]. Schmitt Europe Ltd. also developed a product, LaserCheckTM Inline surface roughness sensor, for measurement of surface qualities based on laser light scatter detection technology. While high sensitivities have been achieved by these methods, only one line or multiple lines on the rods surface are inspected. Thus, it is highly biased and cannot provide enough information for the whole rod surface quality. On the other hand, eddy current sensors have also been widely developed and implemented for surface quality inspections. Despite its non-destructive, simple structure, and fast inspection time, its accuracy can be easily affected by the distance between the sensor, *i.e.* sensing coils, and the rod surface, and the subsurface material properties, e.g. porosity, slag, or even different grain size or hardness [3-5]. Thus, complicated sensor adjustments are needed or large errors could occur. Optical methods, including optical microscopy, light scattering, and machine vision system, have more potential and possibilities for surface quality characterization due to their non-destructive and online inspection capabilities. Although many research works have been demonstrated based on machine vision in areas of touch panel inspection [6], aluminum surface roughness measurement [7], and tile/fabric defect detection [8, 9], these technologies cannot be applied for rod inspection due to 1), the small size of the defects, e.g. nodule diameters are in the range of 1 μm to 90 μm; 2), the high reflective, cylindric, and the wide variety of defects on the rods surface, which are caused by various chemical and mechanical processes in the production line; 3) uneven illumination due to environmental light condition and constructive imperfection; and 4) the short time needed for inspecting the whole rod surface online.

In this paper, a unique automatic vision-based rod surface inspection system with defect detection and differentiation algorithm is presented. The system consists of a LED light, a rod manipulation stage, a high-resolution line scan camera, and a laptop with an integrated software for system control and image processing. Convolutional Neural Network (CNN) is used for image processing to eliminate uneven background intensity. After image processing, data of defect size, count, types, and location are extracted and saved on the PC, allowing users for accessing the surface quality (defects types, sizes, and counts). Experimental results show that, the developed system is capable of detecting defects as small as 25um and has the ability to differentiate nodule, dent, and scratch with a processing time around 4s per rod, which has not been achieved before.

2. System design

2.1. Overall system

The structure of the automatic vision-based rod surface inspection system is shown in figure 1. A high intensity LED light is used for illuminating the rod surface. The robotic manipulation station is motorized with two DC motors to fully automate the inspection; one motor for rod rotation (Motor 1) and the other one for moving the rod along its axial direction (Motor 2). A line scan camera with 4096 pixels and a Field of View (FOV) of 6 cm (around 14.6 um/pixel), which is oriented parallel to the rod axis, is used to capture the surface image of a rod with 22 mm in diameter and 30 cm in length. It is worth mentioning here that higher resolution camera (e.g. 16K resolution) can be used to capture the whole surface in one revolution. A laptop with a self-developed software is used to coordinate the robotic manipulation stage and the camera for image capturing, and control the light indicator based on the rod condition: red light means bad rods and green light means good rods. Image processing and defect extraction are also performed by the integrated software developed in C++.

The working mechanism of the inspection system is illustrated in figure 2. When there is no defect on the rods surface, a mostly dark image will be recorded by the camera since most of the light is reflected, as shown in figure 2 (a). However, if there are defects on the surface (i.e. nodules, dents, or scratches), some light will be scattered toward the camera direction. Thus, a bright area will be recorded in a dark background, as shown in figure 2(b). The size and shape of the bright area is related to the defect size and shape. To be noted that the rod rotation speed was determined based on rod diameter, camera line rate, and FOV. The camera position is adjusted by a 3-axis linear stage and a 2-axis goniometer stage to get the best image quality.

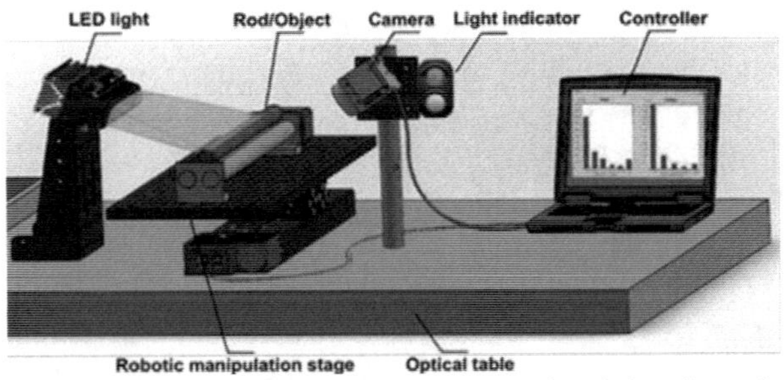

Figure 1. Schematics of the automatic vision-based rod surface inspection system.

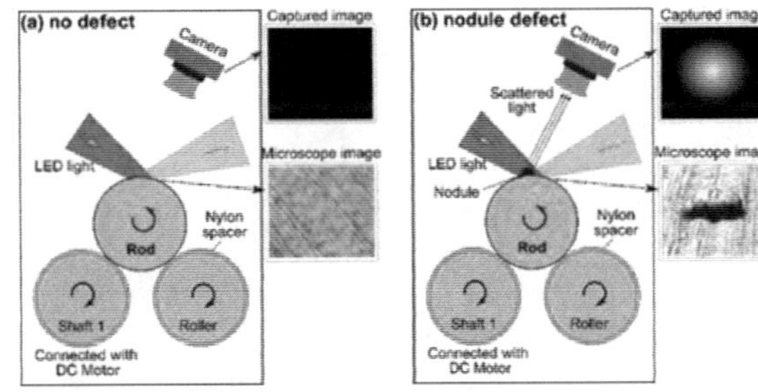

Figure 2. Working mechanism of the rod inspection system.

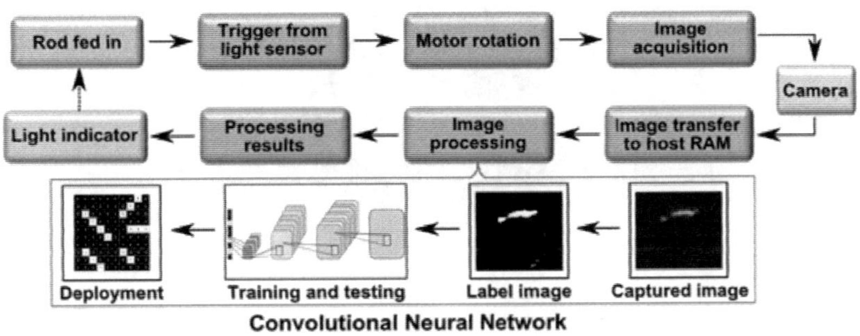

Figure 3. Frame work of the integrated software.

2.2. Image processing framework

The main functions of the integrated software are shown in figure 3. First, the system will initialize for the camera (e.g. line rate, exposure time, frame size, capture time, etc.) and the two motors (e.g. rotation speed, acceleration/deceleration speed, etc.). After that, if the light sensor (in the rod shadow, not shown in figure 1) detected rod presence, Motor 1, which is controlled by the developed software, will start to rotate at the initialized speed. Second, camera will start to record the 360° surface images with the initialized parameters. It is worth to mention here that the line scan camera is calibrated with methods described in [10] for distortion removal. Third, the recorded images (18 frames) are transferred to the allocated PC RAM for processing to get the defect type, size, and count information. To be noted that the FOV is 6 cm in the current design, thus, a total of 5 sections are needed to get the 360° surface images for a 30cm long rod. Motor 2 is used to move the rod from section to section. Finally, the total defect number on the rods surface will be compared with a pre-set threshold. If any defect size overpasses the threshold, the red light will be turned on. If none, the green light will be turned on.

2.3. Convolutional neural networks

Classical image processing is effective when captured image frames have a constant background intensity. When the captured image has varied background intensity, as shown in figure 4(b), and figure 9, binary thresholding becomes unreliable. To overcome this, CNN which has been wide used for speech analysis, object detection and categorization in recent years [11,12], is used to pre-process the captured images. To use CNN, basically, training data sets including input image and label image are needed to train a network/architecture, if the error rate of the output image is less than a threshold, the training is done, and the final architecture is then used in figure 4 (a). Figure 5 gives an overview of the underlying parts of the neural network development. The starting point is the generation of a dataset. After capturing 342 images with a size of 4096 × 262 from different rods, for each image a corresponding label image need to be created. The label images represent the information that the neural network needs to learn, *i.e.* the locations and sizes of defects. For the neural network, an architecture including types and number of layers as well as the structure of their interconnectivities needs to be defined. Then, the dataset – consisting of both captured images and label images – are put into the network during training. In every training step, the output of the network is compared with the label image and the resulting error is projected backwards through the network.

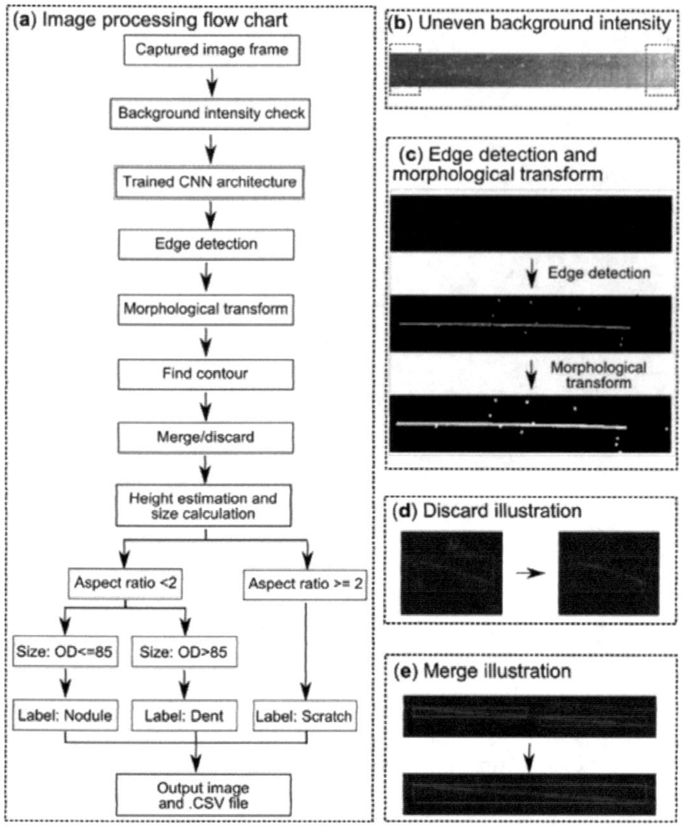

Figure 4. (a) image processing frame work for defect information extraction; (b) illustration of uneven background intensity; (c)- (e) illustration of morphological transform, discard and merge functions.

After that, the weights of the network are adjusted in a way, which reduces the overall error. When training is completed, the trained weights can be used together with the defined architecture to deploy the network in an application, as shown in figure 4(a). Furthermore, the accuracy and performance of the network can be evaluated with a set of test images to see whether the results are satisfying.
Since a lot of defects appear really small on the image, having only a few pixels, the network should make for each pixel a decision if it belongs to a defect or not. For that reason, a binary label image is generated, on which every defect pixel is white and every background pixel is black. How this is done, is shown in the red rectangle in figure 5. As shown in the right part of the red rectangle, the captured image is binary threshold to mark the bright defect clearly. In the left part of the red rectangle, sharp operation is used to increase its contrast before binary threshold. With this step, less bright defects can be found on the resulting binary image with the cost of a very noisy detection of the bright defects. However, the two techniques can be combined to get both advantages of clearly found bright defects and the detection of less bright defects by simply adding the two binary images. Then region of interest (ROI) is used to crop the label image for CNN output training. For the CNN input training data, same ROI on the captured image is cropped then varied its brightness multiplying a random factor between 0.25 and 4, while the corresponding label images remain the same. With this action, the problem of having a dataset with label images, which highly depend on the brightness of their underlying source images, is resolved. Thus, the network can learn to not only rely on brightness, but also on shape and the appearance of the defects.

It should be noted that, as long as the image dimensions are significantly bigger than the size of the biggest filter kernel of the network, the quality of the output is nearly the same for different input dimensions. This fact allows us to crop the images into smaller parts and, by doing so, increases the

IOP Conf. Series: Materials Science and Engineering **999** (2020) 012012 doi:10.1088/1757-899X/999/1/012012

dataset. Since the cropping of a small image part (128 × 128 pixels) is random, not only the number of resulting images can be increased, but also the amount of information in the images. The final dataset has 102600 images with 128 × 128 pixels for training and validation, and 180 images with 4096 × 262 pixels for testing.

To make a pixel-wise prediction using CNN, the output dimensions of the last layer must fit the image input size. Since some of the defects might only have a few pixels, there are no pooling layers in the architecture, because small defects would be lost after down sampling the feature maps. Hence, the size of the feature maps is the same all over the network. In addition to that, there are no fully connected layers neither, to keep the spatial information of the feature maps. All the mentioned constraints lead to an architecture which only consists of convolutional layers followed by ReLU (Rectified Linear Unit) layers as non-linear activation functions.

After generating the dataset and setting up the architecture, training can be started, which is the process of optimizing the weights of the network with respect to a loss function, which calculates the difference between the ground truth represented by the label image and the output image of the network. Since the outputs of our network represent probabilities, it is appropriate to use the Sigmoid-Cross-Entropy-Loss E, which is described in the following formula:

$$E = -\frac{1}{N}\sum_{n=0}^{N}\left[p_n \log \hat{p}_n + (1 - p_n)\log(1 - \hat{p}_n)\right]. \tag{1}$$

where N denotes the number of pixels of the image, p_n the label value of the nth pixel {0,1} and $\hat{p}_n$ the predicted value of the nth pixel. As optimization algorithm, Stochastic Gradient Descent is used. To determine the best architecture, try and error in terms of both structural parameters (number of filters, kernel size, etc.) as well as hyperparameters (learning rate, batch size, etc.) can be used. For architectures with similar performance (error rate of predication), additional method can be used to quantify the quality of the neural network result, which is average of width and height accuracy (ACC), as shown in figure 6.

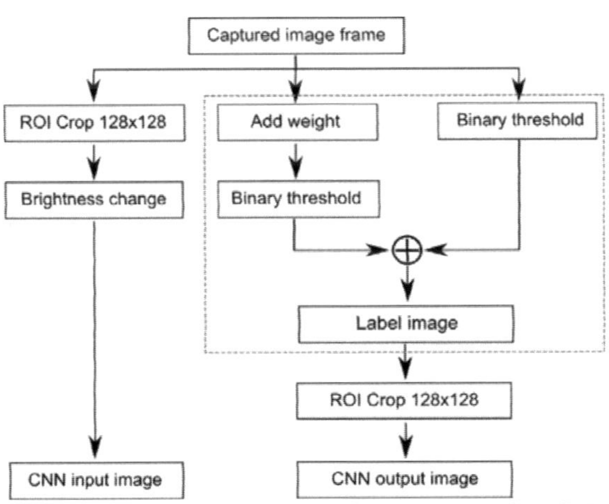

Figure 5. Data preparing frame work for CNN training.

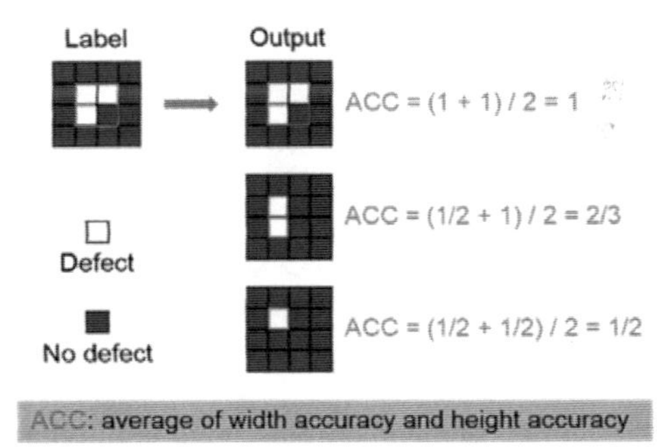

Figure 6. Calculation of defect size accuracy.

Since the final defect size accuracy determined by the bounding boxes after applying the finding contours function, the size of each resulting rectangle of the output image is compared to the corresponding rectangle of the label image. Because nearly no false positives occur (< 0.1%), the lingering ones do not affect the result. On the other hand, missed detections are considered with a size accuracy of zero for the underlying defect. How the size accuracy is calculated in detail, can be found in figure 6. Since most of the defects have a circular shape, we are interested in the accuracy of the diameter prediction. So not the numbers of pixels within the bounding boxes are compared, but their widths and heights (which are the same for the most cases). Both width and height of the bounding box on the output image is compared with the corresponding dimensions of the bounding box of the label image and their accuracies are averaged. This mean value represents the accuracy of the specific detected defect. Then, this value of every single defect is again averaged to calculate the size accuracy for the whole image, and based on this, the best architecture is selected. The finally selected architecture has 5 layers, with 8, 16, 32, 64, and 1 filters per layer. The kernel size for each layer is 15, 9, 7, 5, 3, and the ACC is 85.3% for the whole testing images. To be noted that, ACC as high as 85.3% was obtained with extra layers and kernels, with an expense of much longer processing time.

2.4. Defect classification and calibration

After determined, the CNN model is implemented into the image processing, as shown in figure 4. It is worth mentioning that in this step, labels that have two bright pixels or lie within other labels are discarded or merged together, as shown in figure 4 (d) and (e). Finally, the total defect sizes and numbers are obtained. To classify the defects, aspect ratio and size information are used. If the defect aspect ratio is larger than 3, it will be classified as scratch. However, this cannot be applied to classify nodule and dent because of their similar circular shapes (aspect ratio between 1 and 2.5). To overcome this problem, the size of a total of 774 nodule defects and 90 dent defects were measured under a microscope (Keyence, VHX-600). Their size distribution is shown in figure 7 (a), where nodule sizes range from 2 - 90 μm with a Gaussian distribution, and dent size ranges from 70 μm to 300 μm. Thus, a size threshold of 88 μm was used to classify nodules from dents with an error rate around 1.5%.

Furthermore, the relationship of nodule diameter and its height is measured with a laser confocal microscope (OLYMPUS OLS5000) from Lurie Nanofabrication Facility (LNF) at the University of Michigan. As shown in figure 7 (b), this information is used to estimate nodule height from the detected nodule diameter, as the height information is also a very critical information to determine the surface quality. Figure 7(c) shows the relationship between detected defect pixel number and the defect size measured under the laser confocal microscope, which is used as the calibration curve to get the real defect size from detected defect pixel numbers.

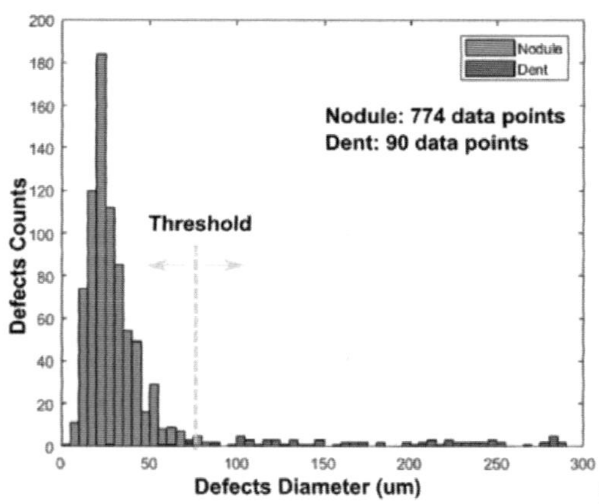

(a)

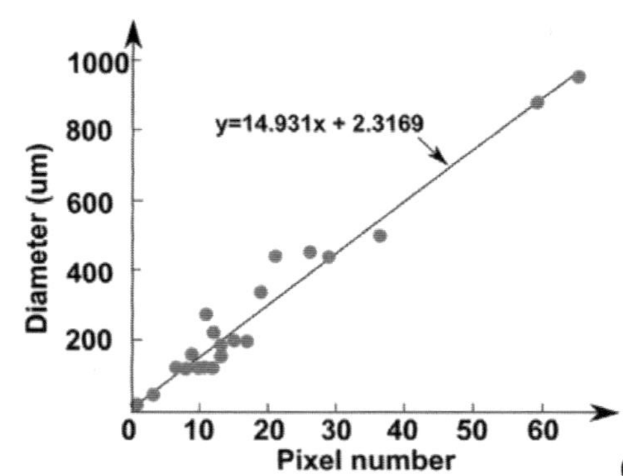

(b)

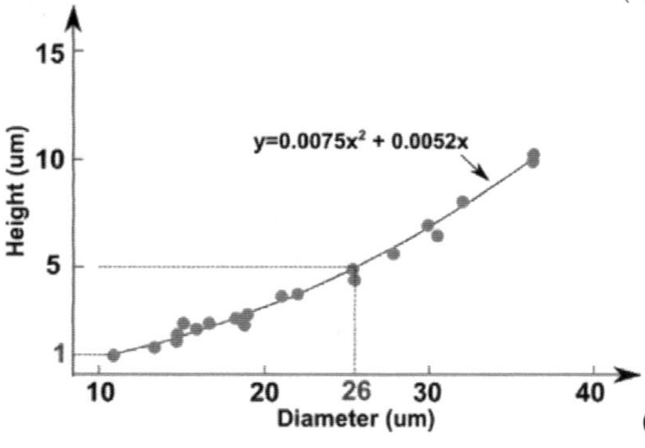

(c)

Figure 7. (a) nodule and dent size distribution and the threshold (88um) that used for classification; (b) relationship between nodule diameter and its height; (c) relationship between actual defect diameter and detected pixel numbers.

3. Experiments and results

Figure 8 shows the setup of the automatic rod inspection system. Two limit switches and one reset switch are implemented for safety purposes. Suspension rods with a diameter of 22 mm were loaded onto the robotic manipulation stage manually. For each section, the image processing and defect classification will be performed after the 360° surface image is captured. The rod condition will then be indicated by the indication light after all the sections are inspected and the defects type, size, and counts are collected. The image processing time for each rod is around 4s. In figure 9, the performance of CNN to overcome the uneven background intensity is compared with conventional binary thresholding method. As can be seen, the CNN effectively removed the uneven background intensity, and correctly detected the defects in the image. Figure 10 shows a sample inspection result. Figure 10 (a) is the image captured from the camera, which is cropped from a 4096 × 262 frame to show the details clearly. Figure 10 (b) shows the processed image with labelling, where green rectangles are the detected nodules, blue rectangles are the detected dents, and red rectangles are the detected scratches. The corresponding microscope images, the captured image, and the labelled image of defects (c), (d), and (e) are also shown in figure 10. The inspection results demonstrated that the current inspection system can effectively detect nodules as small as 25 μm and can differentiate nodules, dents, and scratches with a detection accuracy >90% and sizing accuracy >75%.

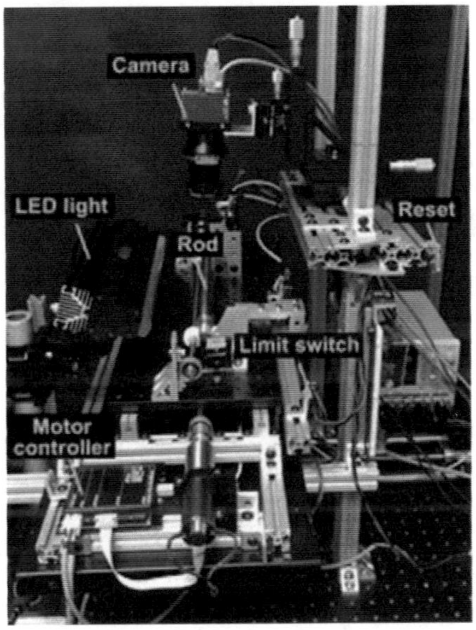

Figure 8. Setup of the automatic rod inspection system.

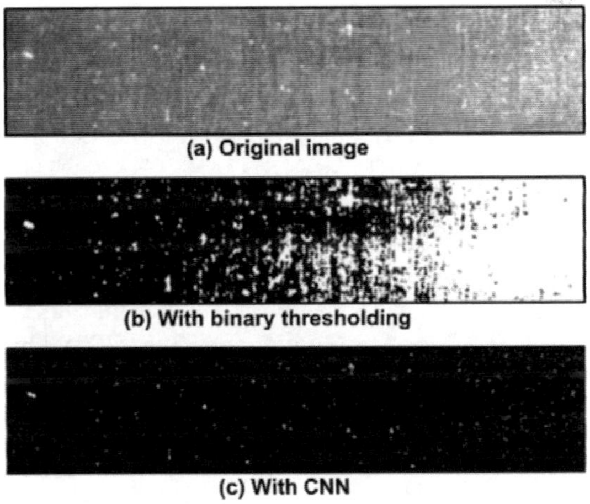

Figure 9. Comparison of image processing results with binary thresholding and CNN.

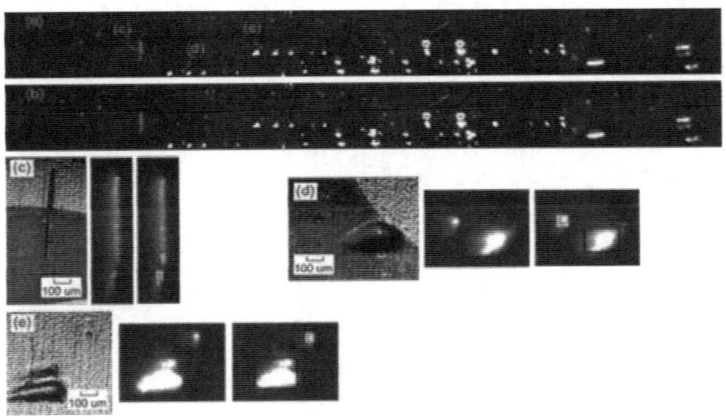

Figure 10. (a) captured raw image of a rod. (b) image after processing and labelling. (c) microscope image of a scratch with 30 μm width and 780 μm length. (d) microscope image of a 30 μm nodule (in red dashed rectangle) and a dent. (e) microscope image of a 25μm nodule (in red dashed rectangle) and two dents.

To be noted that in figure 10 (c), the bottom of the scratch is detected as multiple nodules (green rectangles) as well as some undetected small defects in figure 10 (b). This is because for certain smaller or shallow defects, their pixel intensity in the captured image are low and inconsistent. After CNN processing, these low intensity pixels are omitted in the output binary image, and thus causing this error in the detection image, which can be overcome with functions of edge drawing algorithm (ED), line segment detection (LSD) or Hough Transform techniques [13].

4. Conclusions

Aiming to overcome the disadvantages of the current inspection method for cylindric objects, in this paper, a unique automatic suspension rod inspection system that can detect and differentiate nodules, dents, and scratches is presented. An integrated software is used for LED light control, rod rotation, image capturing, image processing, and indication lights control. For each rod, a total of 90 frames, each with 4096 × 262 pixels were captured and processed. To be noted that, higher resolution camera e.g. 16K resolution, can be used to replace the current 4K camera to get a higher sensitivity, or capture the whole surface image in one revolution. Experiments results shown that, the system can detect defects as small as 25 μm (75% sizing accuracy, and >90% detection accuracy) with a processing time around 4s per rod, which proved the excellent capability of the developed system.

Acknowledgements

Some measurements were partially performed at the University of Michigan Lurie Nanofabrication Facility.

Reference

[1] Wang W and Song Y 2019 Analytical computation method for steady-state stochastic response of a time-delay nonlinear automotive suspension system. *Mech. Syst. Signal Pr., 131*, 434-445
[2] Ismagilov I R, Golenishchev-Kutuzov V A, Kalimullin R I, Migachev S A, and Khasanov A A 2014 Detecting surface and volume defects in metals by the laser-acoustic method. Russ. J. Nondestruct., 50(6), 318-324
[3] Perevertov O, Neslušan M, and Stupakov A 2017 Detection of milled 100Cr6 steel surface by eddy current and incremental permeance methods. *NDT & E Int. 87*, 15-23
[4] Jiao D, Ni L, Zhu X, Zhe J, Zhao Z, Lyu Y, and Liu Z 2019 Measuring gaps using planar inductive sensors based on calculating mutual inductance. *Sensor Actuat. A-Phys.*, 295, 59-69
[5] Han Y, Zhong C, Zhu X, and Zhe J 2018 Online monitoring of dynamic tip clearance of turbine blades in high temperature environments. *Meas. Sci. Technol.*, 29(4), 045102
[6] Chang M, Chen B C, Gabayno J L, and Chen M F 2016 Development of an optical inspection platform for surface defect detection in touch panel glass. *Int. J. Optomechatroni*, 10(2), 63-72
[7] Ghodrati S, Mohseni M, and Kandi S G 2019 Application of image edge detection methods for precise estimation of the standard surface roughness parameters: polypropylene/ethylene-propylene-diene-monomer blend as a case study. *Measurement, 138*, 80-90
[8] Xie L, Xiang X, Xu H, Wang L, Lin L, and Yin G 2020 FFCNN: A Deep Neural Network for Surface Defect Detection of Magnetic Tile. *IEEE Transactions on Industrial Electronics*. DOI: 10.1109/TIE.2020.2982115
[9] Makaremi M, Razmjooy N, and Ramezani M 2018 A new method for detecting texture defects based on modified local binary pattern. *Signal Image and Video P. 12*(7), 1395-1401
[10] Sun B, Zhu J, Yang L, Yang S, and Niu Z 2016 Calibration of line-scan cameras for precision measurement. *Appl. optics*, *55*(25), 6836-6843
[11] Bernal J, Kushibar K, Asfaw D S, Valverde S, Oliver A, Martí R, and Lladó X 2019 Deep convolutional neural networks for brain image analysis on magnetic resonance imaging: a review. *Artifi*. Intell. Med. *95*, 64-81
[12] Fahad S A, and Yahya A E 2018 July Inflectional review of deep learning on natural language processing. In *2018 International Conference on Smart Computing and Electronic Enterprise (ICSCEE)* (pp. 1-4). IEEE
[13] Liu Y, Xie Z, and Liu H 2019 LB-LSD: A length-based line segment detector for real-time applications. *Pattern Recognit. Lett.* 128, 247-254.

IOP Conf. Series: Materials Science and Engineering **999** (2020) 012013 doi:10.1088/1757-899X/999/1/012013

Waste Elimination Model Based on Lean Manufacturing and Lean Maintenance to Increase Efficiency in the Manufacturing Industry

P Aucasime-Gonzales[1], S Tremolada-Cruz[1], P Chavez-Soriano[1], F Dominguez[2] and C Raymundo[3]

[1] Ingenieria Industrial, Universidad Peruana de Ciencias Aplicadas, Lima, Peru.
[2] Escuela Superior de Ingenieria Informatica, Universidad Rey Juan Carlos, Madrid, Spain.
[3] Direccion de Investigacion, Universidad Peruana de Ciencias Aplicadas, Lima, Peru.

Email: U201512412@upc.edu.pe; U201511694@upc.edu.pe; pedro.chavez@upc.edu.pe; carlos.raymundo@upc.edu.pe

Abstract. This article focuses on the low efficiency of a production process in a manufacturing company. The study aims to integrate and model different tools with greater efficiency to avoid high maintenance costs of production equipment and high implementation costs. Thus, this article presents a waste reduction model intended to reduce setup time and develop an efficient maintenance management system to increase overall equipment effectiveness (OEE). We validated the model in a Peruvian tire manufacturing company and achieved positive outcomes such as a 13% increase in the OEE score and a 22.5% setup time reduction. All this entailed an increase in the component changeover capacity, the outsourcing of 37 activities, and an 80% increase in the knowledge index of the operators involved in the proposal.

1. Introduction

The rubber and plastic sector in Peru record a negative production volume variance, while other sectors present a positive production volume variance [1]. Therefore, we analysed the tire manufacturing process, where we found various issues such as low productivity, quality problems, and production machine downtime; the latter has the greatest impact because it affects equipment performance [2].

The most critical process of the tire industry is vulcanization due to the curing press defects, which makes the process inefficient because the press efficiency is directly affected. In several cases, this results in failure to produce more than 10% of the daily production plan [2].

Competitive advantage is a crucial factor for survival in the Peruvian manufacturing industry. Thus, we intended to integrate various engineering tools in a model that is more efficient than earlier ones, to avoid high implementation and maintenance costs. For this reason, and given the research motivation, we proposed a waste reduction model based on lean manufacturing and lean maintenance to increase efficiency in the manufacturing industry.

2. State of the Art

2.1. Lean manufacturing implemented in manufacturing industries

The authors state that companies implement lean manufacturing techniques to reduce production process waste such as changeover time. In several cases, successful tool implementation improved the overall equipment effectiveness (OEE) score by approximately 4%; in one case, the changeover time was reduced by 40%, while in another case, a 87% reduction of setup time was achieved in a manufacturing process [3][4]. Other success stories related to setup time reduction show decreases of 50% and 64% as a result of task outsourcing, such that work is done based on the early preparation principle. [5][6].

2.2. Lean maintenance implemented in manufacturing industries

Another lean philosophy tool is lean maintenance, referring to the implementation of lean tools for maintenance management, in manufacturing machinery that presents high levels of waste, such as machine stoppages. For example, the implementation of lean maintenance in textile production resulted in a downtime reduction by 26% and an increase in machine availability by 1.7%. In other scenarios, the positive impact is also shown by an improved OEE indicator, which increased by 5%–7% [7][8][9]. Similarly, the implementation results show an increase in the number of products that can be manufactured (by approximately 31%), a 30% reduction in machine downtime, an average operational efficiency of 67.1%, and an increase in equipment availability by up to 78.9%. Moreover, as mentioned above, the greatest benefit in operational performance indicators translates into an improved OEE score, which in a couple of scenarios could be increased from 45.36% to 60.38% and which averaged 51.5% [2].

3. Contribution

3.1. Proposed Model

As part of the contribution, figure 1 presents a waste reduction model based on lean manufacturing and lean maintenance. It integrates several components that have been comprehensively implemented in the scenario. The model has four phases. The first component is staff because the first phase focuses on its training and education in relation to the methodology and benefits of the whole model. The staff component plays a crucial role because based on their experience and continuous learning, workers will manage to use various tools to improve their current work method and achieve their goals [10].

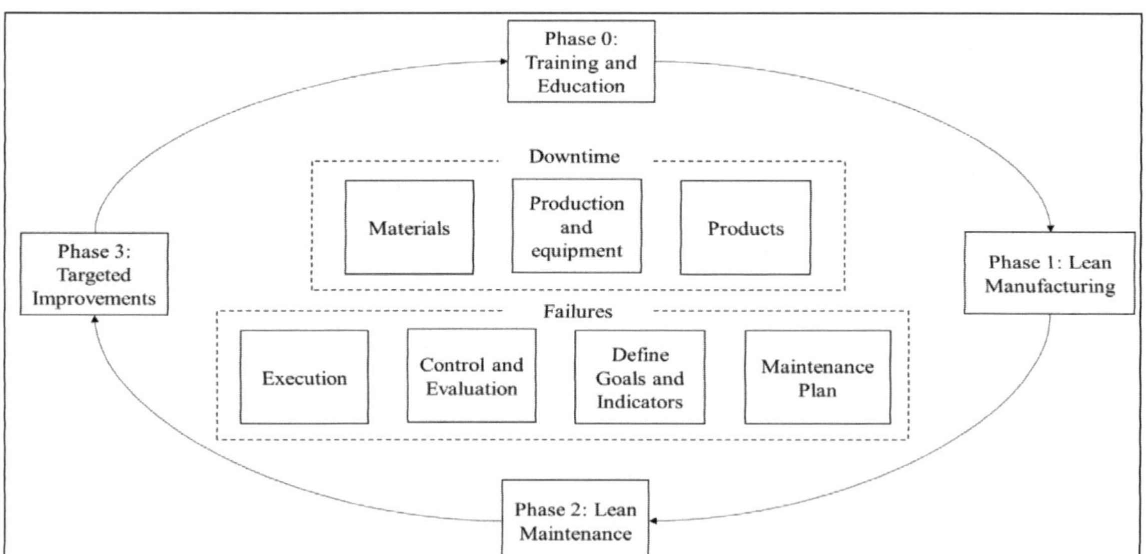

Figure 1. Waste reduction model

The second component is waste; production equipment downtime, which directly affects the manufacturing company products, causes various problems such as production noncompliance, defects, and loss of man-hours and machine-hours. Thus, companies seek to minimize or, at best, eliminate waste by using lean manufacturing techniques. The third and last component consists of the waste lean defects, involving production machines and their maintenance. This is an important factor for the elimination of this type of waste, from evaluation to building a maintenance plan, focusing on lean maintenance. Another goal is to sustain the model over time and continuously improve it to achieve better performance.

3.2. Proposed method

The waste elimination model was implemented using the *Single Minute Exchange of Dies* (SMED) and autonomous maintenance pillar of total productive maintenance. Methodology is shown in figure 2.

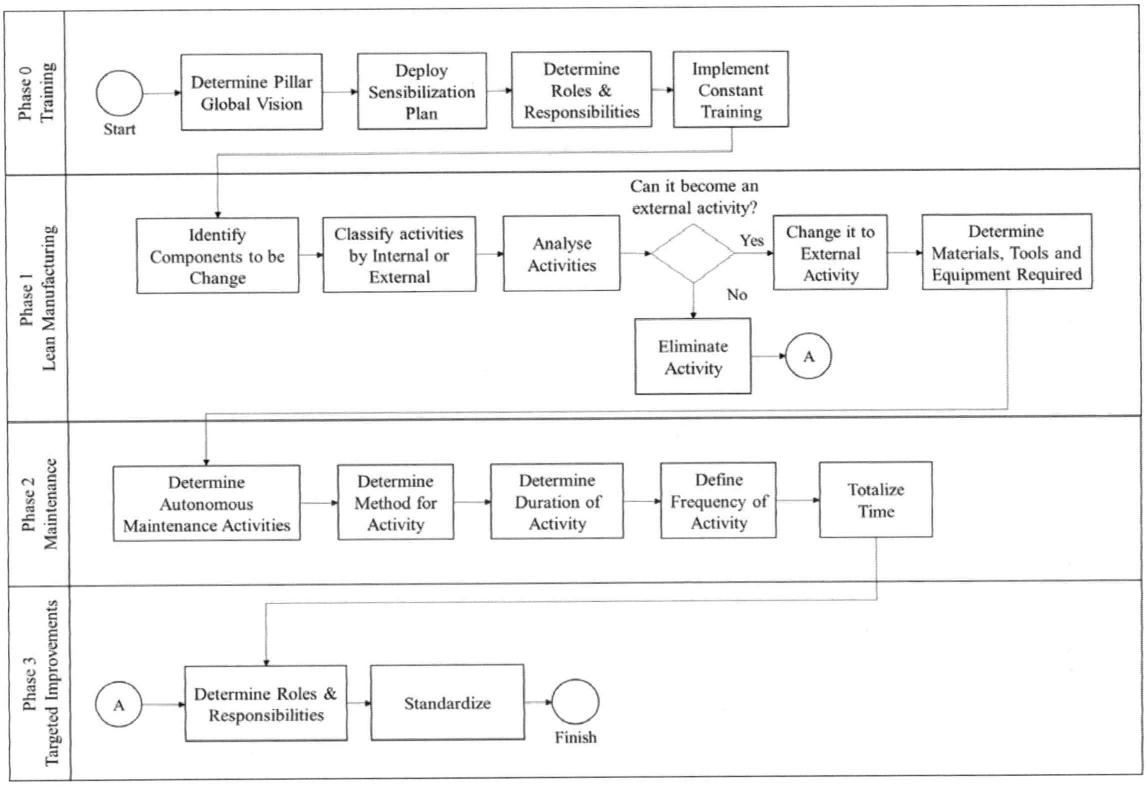

Figure 2. Implementation flowchart of the proposal

3.3. Indicators.

Below are the main metrics related to the primary model goal.

3.3.1. Set-up time reduction. Goal is to reduce setup times by at least 11% and its interpretation is to show the changeover time reduction in a process.

$$\frac{(End\ time\ -\ Initial\ time)}{Initial\ time} \times 100\% \tag{1}$$

3.3.2. Overall equipment effectiveness (OEE). The indicators presented will be used to measure the proposed model performance and show the achievement percentage of the proposed goals. Goal is to increase efficiency by at least 4% and its interpretation is to increase overall equipment performance.

$$Availability\% \times Performance\% \times Quality\% \tag{2}$$

4. Validation

4.1. Case study

The proposed model was implemented and validated in a tire manufacturing company in Lima, Peru, specifically for the vulcanization process, that is, the last process of tire manufacturing and in a radial tire vulcanizing press. Figure 3 shows one of the vulcanizing presses as part of the scenario where the case study has been carried out.

Figure 3. Vulcanization press

4.2. Initial diagnosis

Regarding the initial diagnosis of the case study, we obtained the following results for the vulcanizing press used for the pilot plan.

- The current changeover time is 37.8 min. for moulds and 27.4 min. for bladders.
- 37 activities were carried out in a wrong manner in the mould and bladder changeover activity.
- The initial OEE score in the pilot press was 55.7%

The press in question had an OEE score of 55.7%, which is well below the world-class OEE. Further, a time study showed that the press bladder and mould changeover times exceed, on average, 33 minutes. The time taken is very long; thus, the process is inefficient and entails a current changeover capacity of 4 moulds and 6 bladders per day, that is, only 57% of the operator workload.

4.3. Contribution to the scenario

We developed the implementation method for the waste elimination model starting from the ongoing training stage given to the pilot press operators and then implemented SMED in the bladder and mould changeover activities to achieve a work method that enhances the vulcanization process efficiency. Finally, we implemented autonomous maintenance to improve machine operation by reducing downtime.

All the changeover activities to be performed must be mapped along with their execution time, to analyse them and to choose the activities to be outsourced, as implemented in the tool shown in table 1. The goal is to reduce high setup times, performing some of the activities while the press continues to operate, without the need to stop it, which wastes tire vulcanization time.

Table 1. Classification of activities

Activity	Task	Task Duration (min.)	Activity Duration (min.)	Type	Proposal
Mould Change	Loosen bolts of the previous mould	2.70	36.81	Internal	External
	Bring the crane with tools to press	1.32		Internal	
	Lift the press head	0.64		Internal	

Likewise, because of the constant and prolonged press stoppages in the tool shown in table 2, we presented the autonomous maintenance standards to achieve effective maintenance management.

Table 2. Autonomous maintenance standards

Inspection / Lubrication / Cleaning Standards			
Area: Vulcanization			
Activity	*Method*	*Time (min)*	*Frequency (weeks)*
Sean check and cleaning	Cleaning	5	3
Safety bar check	Inspection	3.5	4
Main gear oil change	Lubrication	8	12

Thus, using the proposed tools and based on the initial press diagnosis indicators and the outcomes achieved, we made a comparison, as shown in table 3.

Table 3. Comparing initial vs final outcomes

Title	Outcomes before implementation	Outcomes after implementation
Set-up time	64 min	50 min
Number of activities performed in a wrong manner	All the activities were carried out internally	37 outsourced activities
New pilot press OEE indicator	55.70%	68.8%
Previous Knowledge index	1/4	4/4

As shown in table 3, the setup time of the bladder and mould components was reduced by 14 minutes, which increased the component changeover capacity by 7 bladders and 2 moulds per day. This resulted from SMED implementation and a work method improvement based on the tools used to outsource 37 activities that could be carried out when the press is in operation. Likewise, the press OEE indicator was increased by 13%, mainly using autonomous maintenance based on tools such as the cleaning, inspection, and lubrication register, which detailed the related activities, as well as the manner and frequency in which they must be carried out.

5. Conclusions

By implementing the improvement model, we achieved an OEE increase of 13%, a 22.5% reduction in setup times, an increase in the component changeover capacity by 7 bladders and 2 moulds and an 80% increase in the knowledge index.

The tools play a significant role in the proposal implementation because they are used to conduct the proposal components step-by-step, and they support all stakeholders in the efficient development of the proposal.

The commitment of the company's employees, from top management to operators, is of great importance, because they are the ones to develop the proposal and have great influence on the techniques used.

References

[1] SNI, "*Importancia de la Industria*", 2018

[2] F. Saleem, S. Nisar, M. A. Khan, S. Z. Khan, and M. A. Sheikh, "Overall equipment effectiveness of tyre curing press: A case study", J. Qual. Maint. Eng., 2017

[3] J. Singh, H. Singh, and I. Singh, "SMED for quick changeover in manufacturing industry – a case study", Benchmarking, vol. 25, no. 7, pp. 2065–2088, Oct. 2018

[4] M. Braglia, M. Frosolini, and M. Gallo, "Enhancing SMED: Changeover out of Machine Evaluation Technique to implement the duplication strategy", Prod. Plan. Control, vol. 27, no. 4, pp. 328–342, 2016

[5] Karwasz and P. Chabowski, "Productivity increase through reduced changeover time", J. Mach. Eng., vol. 16, no. 2, 2016

[6] K. Antosz and A. Pacana, "Comparative analysis of the implementation of the SMED method on selected production stands", Teh. Vjesn., vol. 25, pp. 276–282, 2018

[7] Thawkar, V. Deshpande, and P. Tambe, "A reliability centred maintenance approach for assessing the impact of maintenance for availability improvement of carding machine", Int. J. Process Manag. Benchmarking, vol. 8, no. 3, p. 318, May 2018

[8] Henríquez-Alvarado, F., Luque-Ojeda, V., Macassi-Jauregui, I., Alvarez, J. M., & Raymundo-Ibañez, C. (2019). Process optimization using lean manufacturing to reduce downtime: Case study of a manufacturing SME in peru. Paper presented at the ACM International Conference Proceeding Series, 261-265. doi:10.1145/3364335.3364383

[9] T. Ylipää, A. Skoogh, J. Bokrantz, and M. Gopalakrishnan, "Identification of maintenance improvement potential using OEE assessment", Int. J. Product. Perform. Manag., 2017

[10] Y. Cano, G. Quispe, H. Chavez, N. Mamani-Macedo, C. Raymundo-Ibañez, F. Dominguez (2020) Occupational Health and Safety Management Model for Mining Contracts. Human Interaction, Emerging Technologies and Future Applications II. IHIET 2020. Advances in Intelligent Systems and Computing, vol 1152. Springer, Cham. DOI: https://doi.org/10.1007/978-3-030-44267-5_74

IOP Conf. Series: Materials Science and Engineering **999** (2020) 012014 doi:10.1088/1757-899X/999/1/012014

Waste Reduction with Lean Manufacturing Model in an Alpaca Wool Workshop

J Cristobal[1], G Quispe[2], F Dominguez[3], G Zapata[4] and C Raymundo[4]

[1] Facultad de Ingenieria, Universidad Continental, Huancayo, Peru
[2] Universidad Nacional Autonoma Altoandina de Tarma, Junin, Peru
[3] Escuela Superior de Ingenieria Informatica, Universidad Rey Juan Carlos, Madrid, Spain.
[4] Direccion de Investigacion, Universidad Peruana de Ciencias Aplicadas, Lima, Peru

Email: 72674845@continental.edu.pe; gquispe@continental.edu.pe; carlos.raymundo@upc.edu.pe

Abstract. This paper proposes the design of a quasi-experimental time-series model comprising lean manufacturing tools—such as the value stream mapping and 5S's—with the aim of reducing the level of waste and losses in a micro-company dedicated to the production of Alpaca wool clothing in Junín, Perú. The model implementation managed to reduce the level of waste by 7%, increase the efficiency in material consumption, and reduce the cycle time and total supply time. The implementation of the model allowed reducing the level of waste in the production process and creating an organizational culture based on order, cleanliness, discipline, and teamwork.

1. Introduction

The Peruvian textile industry represents 1.3% of the GDP and 8.9% of the country's productive sector [1]. This industry includes companies that work with Alpaca fiber, with a large percentage of their production being destined for export due to the quality of their yarn. Thus, in 2018, exports grew by 22.1% with respect to the previous period, equivalent to more than USD 12,000,000.00 [2].

As Alpaca yarn accounts for 70% of the total production costs, it is vital to keep the percentage of waste or excess material to a minimum. The variation in waste and excess material directly impacts the final quality of the products and profits of the organization [3]. Nowadays, the maximum limit of waste accepted in the Alpaca fiber sector is 4.4% [4].

The importance of reducing the level of waste within the textile industry has been recognized by several authors. Extensive research has been conducted regarding the same, of which the umbrella model stands out. This model has managed to reduce waste resulting from overproduction, reprocessing, non-value-added work, and non-compliance in cost movement [5]. Another study verified the use of lean manufacturing tools for reducing the total supply time and the waste generated, with minimal investment in machinery, training, and technology [6]. In addition, it has been demonstrated that the integration of Kanban tools and 5S's manages to reduce not only the total supply time and flow of information but also defects [7] [8]. Finally, it has been shown that its use promotes a work environment that motivates the worker to achieve sustainability in the organization and increase productivity [9].

Thus, a model that, supported by the integration of VSM and 5S's, allows for a reduction in the level of waste throughout the production process is desired. The objective is not only to contribute to

IOP Conf. Series: Materials Science and Engineering **999** (2020) 012014 doi:10.1088/1757-899X/999/1/012014

increasing the profits of industrial micro-enterprises, by reducing waste, but also to create sustainable work environments that promote continuous improvement.

2. State of the Art

2.1. Application of lean manufacturing in the textile industry

The implementation of Lean manufacturing tools in the textile industry aims to reduce waste and increase the level of customer satisfaction. Therefore, a scheme was developed with the following necessary steps for waste reduction: first, reduce the response time; then, allocate the materials and resources necessary for production; and finally, execute production according to demand, seeking cross-cutting management of information.

The results showed a reduction of 80% and an index of 75% of workers informed and trained for the design execution. However, the time required for its implementation is approximately one year. In addition, other tools are needed for a successful implementation [10].

Some other studies proposed the joint application of Kanban, 5S's', and PDCA cycle to reduce downtime, information flow, and through the continuous improvement cycle, defects and waste. The results demonstrated an 84% increase in the organization's income and a 25% reduction in the level of defects, reducing the associated production costs [7].

Likewise, it has been demonstrated that a Lean manufacturing model formed of three tools—layout redesign, source quality, and line balance—can achieve similar results. Thus, its implementation shows a 96.7% reduction in waste and 43% reduction in the total delivery time [6].

3. Methodology

As there is no control over all the variables that are part of the research, a quasi-experimental model was developed. This model requires periodic measurements on a group of Y variables, described in table 1, before and after introducing the X improvements, to determine the effect of the latter on the variables described [11]. The implementation stages are described in figure 1.

Table 1. Description of variables

Variable	Description
Y_1	Raw material waste and loss
Y_2	Defective product
Y_3	Production cycle time
Y_4	Total supply time
Y_5	Added value activities

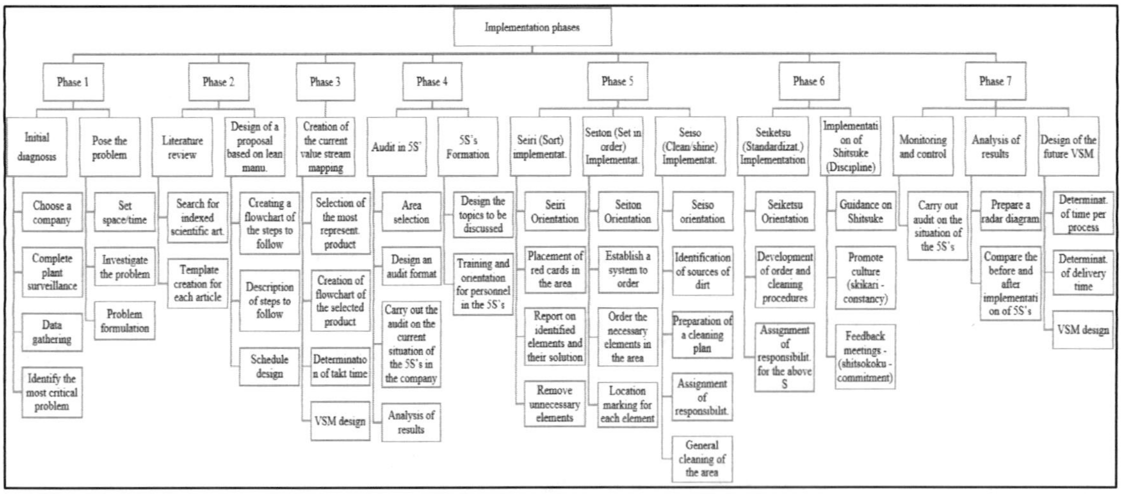

Figure 1. Implementation Design

3.1. Phase 1. Initial diagnosis and problem statement
Select the company and carry out the initial diagnosis to identify the main problems and, after analysis, define their causes. Finally, the scope of the project is defined in terms of space and time, the problem is investigated in depth for its solution and finally the scope of the problem is formulated.

3.2. Phase 2. Bibliographic sources review and design solution proposal
Compare and contrast previous results from lean methodology used in previous researches focused on both waste reduction and its implementation methods. Finally, the solution model is proposed and the implementation schedule is made official.

3.3. Phase 3. Set up Value Stream Mapping
Determine the product or product family most relevant to the organization. Identify the materials and information flow in the selected process. Determine the takt time and the activities with the highest level of waste.

3.4. Phase 4. Carry out a 5S´s audit and form working teams
Select the area, perform the assessment formats and carry out the 5S's audit to know the current state of the organization. Subsequently, train and guide workers in the implementation of 5S's.

3.5. Phase 5

3.5.1. Sort (Seiri). After workers were trained in the element classification in the working area, unnecessary elements were separated using visual control labels, called "red cards." The name of the object, its classification, the reason for removing it, the type of action to take, and observations were recorded. All useless items were removed or relocated.

3.5.2. Set in order (Seiton). After removing the unnecessary objects, the workers were trained in the order of the necessary tools and elements. Workstations were arranged considering the criteria described in table 2 [12].

Table 2. Elements' location criteria

Frequency of use	Location
Hourly	Locate it beside the person
Several times a day	Locate it near the person
Several times a week	Locate it in the workspace
A few times a month	Locate in a labeled place easy to reach

In addition, the equipment, tools, and materials were labeled and located according to their use.

3.5.3. Shine (Seiso). The staff were trained on cleaning the workspace. Subsequently, the sources of dirt were identified and a cleaning plan for assigning responsibilities was prepared. Deep cleaning of tools and machinery for optimal operation was also carried out.

3.6. Phase 6

3.6.1. Standardize (Seiketsu). The personnel were trained in the standardization of activities and procedures and standards were established to eliminate the previously identified the dirt sources at the workspace. Responsible parties were assigned and a schedule was prepared to carry out the specific cleaning.

3.6.2. Sustain (Shitsuke). The workers were trained on tool implementation. Due to the inherent difficulty in measuring this step, given its close relation with the organizational culture and employee behavior, two more steps were implemented: consistency (shikari) and commitment (shitsokoku). The first seeks to ensure that workers comply with the steps previously implemented through periodic visits to the workspace, mainly to the activities identified as having the greatest waste. Regarding commitment, it was obtained progressively, holding periodic meetings to receive suggestions and observations from the workers.

3.7. Phase 7
A final audit was carried out to verify compliance with the tool. The results were then analyzed and the impact generated on the level of waste in each activity was measured, using a radar chart. Finally, a future VSM was elaborated to evaluate the impact of the implementation in time per the process and delivery time.

4. Validation

4.1. Initial diagnosis
The jerseys were identified as the family of products with the highest seasonal demand for the company, representing 21.32% of the total production volume. This product was established as the most relevant for the organization, developing its value stream mapping.

The percentage of waste generated in the jersey production was analyzed considering the models with the highest demand. It was found that, for the production of a sample total of 200 units, comprising 10 units per design, the total weight of the waste was 16.24 kg, which represents 11.67% of the total material entered.

For the development of VSM, the takt time was first determined as 38.64 min/unit. Subsequently, the material and information flow were established to draw the current process of jersey production, as shown in figure 2. A total cycle time of 41.6 min and a total supply time of 3.44 days were obtained. This clearly shows that the supply time was greater than the takt time, which resulted in delays in deliveries.

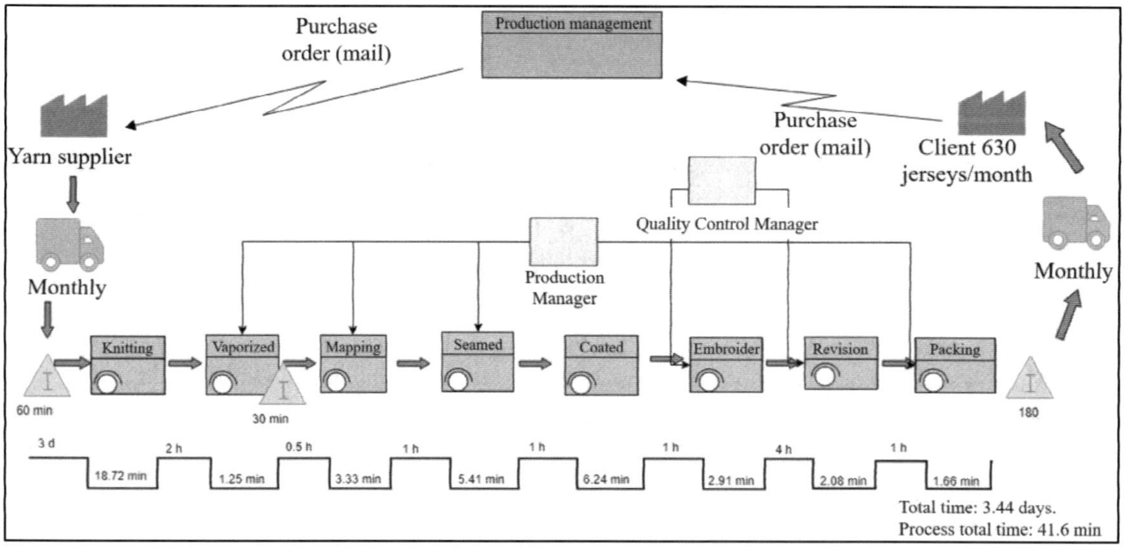

Figure 2. Initial VSM of jerseys in the company

An audit was conducted to determine the level of compliance with 5'S methodology in the workspace. The results showed the little commitment and culture focused on order and cleanliness, obtaining only 32% compliance with the 5S's methodology.

4.2. Implementation

The workers were trained in the objectives and benefits of the project, as well as in the steps for the 5S's tool execution. The implementation was conducted in the following steps:

4.2.1. Sort. Red cards were placed on unnecessary elements in the workspace. As shown in table 3, 103 elements were found, and were relocated to other areas. The item removal was carried out within two days.

Table 3. Actions taken on unnecessary elements

Element	Number	Classification	Action
Thread	22	Raw Material	Relocate
Plastic lid	7	Container	Eliminate / dispose
Carboard box	12	Container	Recycle
Mannequin	2	Equipment	Relocate
Tub	1	Equipment	Eliminate / dispose
Rags	18	WIP	Relocate
Radio	2	Equipment	Repair
Scissors	8	Tool	Repair
Notchers	6	Tool	Eliminate / dispose
Needles	25	Tool	Eliminate / dispose

4.2.2. Order. A specific location was established for each element necessary for production and containers were labeled to facilitate the identification of each element.

4.2.3. Clean. The main sources of dirt were identified in the activities that generated maximum waste. The activities that generated the maximum lint were identified to make the machines to dirty faster and break down more frequently. Walls, floors, machines, shelves, and tools were thoroughly cleaned (See figure 3 and figure 4).

Figure 3. Object labeling at the workspace.

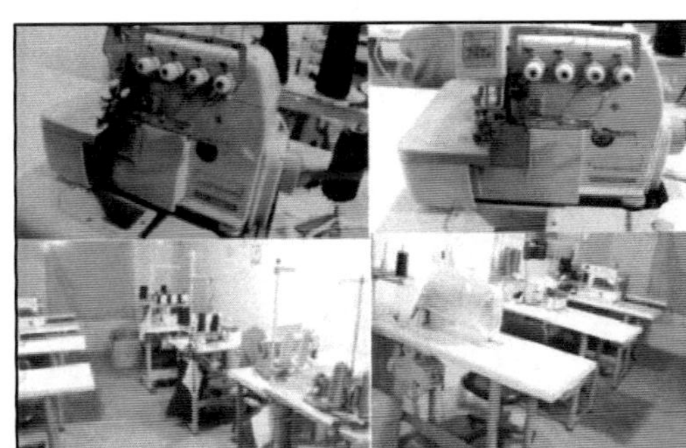

Figure 4. Cleaning of machines and equipment

4.2.4. Standardize. To maintain proper compliance with the implemented steps, a cleaning schedule was developed for the area and responsible parties were appointed.

4.2.5. Sustain. Work was done to reinforce the organizational culture and discipline of the staff.

4.2.6. Consistency. A schedule of periodic visits was set up to monitor and reinforce the order and cleaning habits of the personnel.

4.2.7. Commitment. A schedule of meetings with plant personnel was established to encourage them to continue with good performance and receive feedback on the current situation, as well as opportunities for improvement.

4.3. Results
As shown in figure 5 and 6, after carrying out the final audit, the percentage of compliance with the implementation from 5S's was increased by 84%. Furthermore, the percentage of waste was reduced from 11.67% to 5.00%, with a total waste of 5.82 kg for the production of 200 units of jerseys divided into 20 models.

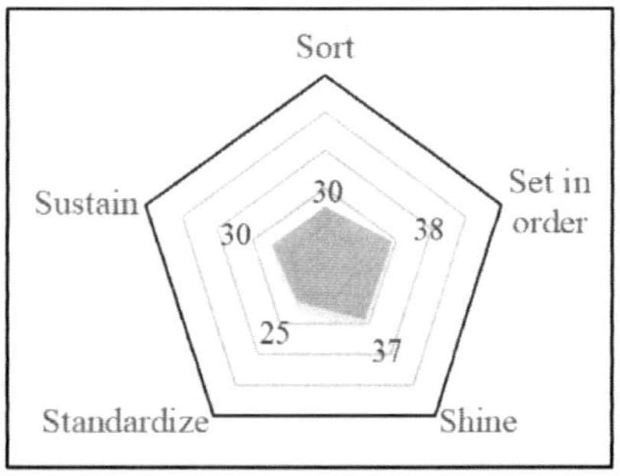

Figure 5. 5S's initial audit radar chart.

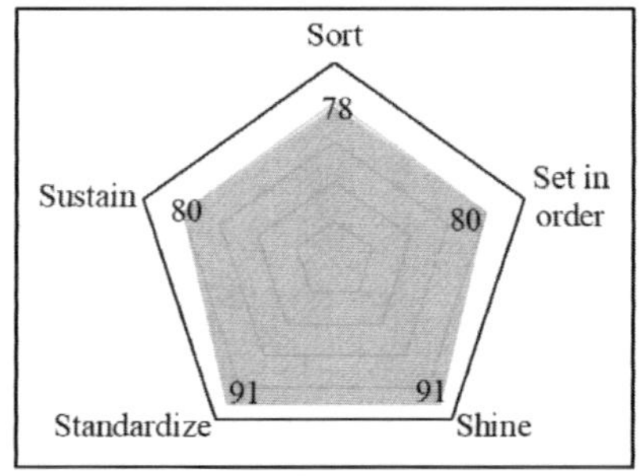

Figure 6. 5S's radar chart after tool implementation.

The Takt Time was also determined taking into account the most recent:

$$Takt\ Time = \frac{Production\ time\ availablee}{average\ demand\ per\ customer} = \frac{12\ 480 \frac{minutes}{month}}{323 \frac{sweater}{month}} = 38.64 \frac{minutes}{sweater} \quad (1)$$

Subsequently, the future VSM was prepared—as shown in Fig. 7— to analyze the impact on material flow in the process. This evidences that the cycle time was 38.6 min, with a total delivery time of 2 days. These times are less than the initial values, allowing the on-time delivery of the products, as it is less than the takt time of 38.64 min/jersey.

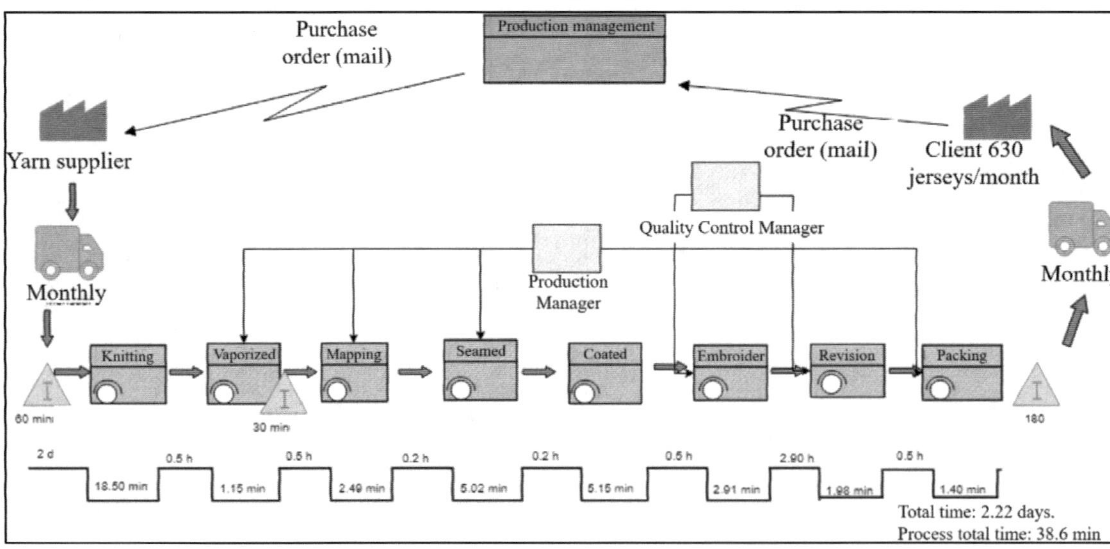

Figure 7. Future VSM for the jersey line production.

5. Conclusions
The research helped to determine the jersey as the most relevant product family, as it represents 21.32% of the total monthly production in the season.

The company's current VSM revealed that there is a cycle time of 41.6 min per garment and a delivery time of 3.44 days; these times are greater than that required to satisfy the demand (takt time = 38.64 min), which causes the late delivery of the orders. A positive result was obtained with the application of the 5'S methodology, as it reduced the level of waste by 7%, besides reducing the cycle time and delivery time, and therefore, the efficiency of the system also improved.

We conclude that lean manufacturing tools reduce the level of waste generated in production and have a positive impact in the workplace. This is because they promote the creation of a discipline-based culture that allows for a cleaner work environment, and encourage the staff to practice teamwork.

References

[1] García JM 2017 Research study of the textile and clothing sector. Ministry of Production [online] Available from: http://ogeiee.produce.gob.pe/index.php/oee-documentos-publicaciones/documentos-de-trabajo/item/725-textil-y-confecciones

[2] Mincetur 2018 Peru: Alpaca textile exports grew 48% in the first half. Peru Retail [online].

[3] Socconini L 2008 The foundations for world-class leadership and culture. In: Lean manufacturing step by step: the Japanese business management system that revolutionized manufacturing and services. ISBN 9789700919324. ISBN 9789700919324

[4] Sáenz RL 2013 Proposal to reduce waste in the garment sector in the textile industry. Institutional Repository - UNI [online]. Available from: http://cybertesis.uni.edu.pe/handle/uni/7049

[5] Cueto G, Caldas Y, Viacava G, Quiroz, J And Alvarez J 2020 Waste Reduction Model in a Small Clothing Company-Umbrella Model. In. ISBN 9783030256289. ISBN 9783030256289

[6] Obeidat M, Al-Aomar R and PEI ZJ 2014 Lean manufacturing implementation in the sewing industry. Journal of Enterprise Transformation. 4 (2), p. 151–171. DOI 10.1080 / 19488289.2014.890980.

[7] Andrade Y, Cardenas L, Viacava G, Raymundo C And Dominguez F 2019 Lean manufacturing model for the reduction of production times and reduction of the returns of defective items in textile industry [online]. Springer International Publishing. ISBN 9783030204433. Available from: http://dx.doi.org/10.1007/978-3-030-20444-0_39ISBN 9783030204433.

[8] Becerra, A., Villanueva, A., Núñez, V., Raymundo, C., Dominguez, F., "Lean manufacturing model in a make to order environment in the printing sector in Peru", Advances in Intelligent Systems and Computing, 971, pp. 100-110.

[9] Laura M, C Alvez A And Celina L 2016 Imece2012-89048 Design of a Lean Methodology for an Ergonomic and Sustainable. Proceedings of the ASME 2012 International Mechanical Engineering Congress & Exposition IMECE 2012. P. 1–10.

[10] Cortez C, Di Laura N, Viacava G, Raymundo C, Dominguez F 2020 Lean Manufacturing Model Based on Knowledge Management to Increase Compliance in the Production Process in Peruvian SMEs in the Textile Garment Sector Advances in Intelligent Systems and Computing, 961, pp. 103-111. DOI: 10.1007/978-3-030-20154-8_10

[11] Valderrama S Steps to evaluate scientific research projects and thesis. Lima, 2007. ISBN 978-9972-38-041-9. ISBN 978-9972-38-041-9

[12] Villaseñor, A. Y Galindo, E. System 5 Ss: Implementation Guide. México : AUTOR-EDITOR, 2010. ISBN 9786070502545. ISBN 9786070502545

Study on the capacity performance of journal bearings with different spiral groove structures by CFD method

Le Zhang[1], Qi Xin[1], Bo Wu[1], Ning Zhong[1], Liyue Tu[1], Hulin Li[2,*]

[1]Chongqing Yuejin Machinery Co., Ltd., No.1 Huagong RD, Yongchuan District, Chongqing 402169, PR China
[2] School of Design, Shanghai Jiao Tong University, 800 Dongchuan RD, Minhang District, Shanghai 200240, PR China
[*]Corresponding author: E-mail address: lhl666@sjtu.edu.cn (Hulin Li)

Abstract. This paper uses the computational fluid dynamics (CFD) method to evaluate the capacity performance of journal bearings with different spiral groove shapes. A detail discussion is conducted to compare the capacity performance of spiral groove journal bearings with different spiral groove depths, widths, numbers, and pitches. For journal bearings with different spiral groove structures, the variation of the load carrying capacity and the oil film pressure with different attitude angles is illustrated. The influence of the spiral groove structures on the pressure distribution of journal bearings is investigated. The presenet research can provide techinal support and references to the design and the capacity performance of journal bearings with the spiral grooves.

Keywords: Journal bearings; Spiral groove; Load carrying capacity; Structure design

1. Introduction

Journal bearings are widely used in rotating machinery due to their stable operation, good vibration resistance, high bearing capacity. Generally, the bearing structure affects the bearing capacity performance. With the development of industry, the demand for journal bearings with high capacity performance urgently needs to be met. It is necessary to design reasonable structures to improved the bearing capacity performance.

Sing et al. [1] found that a reasonable groove shape can greatly improve the bearing performance. Sharm et al. [2-3] compared the capacity performance of bearings with different groove shapes and the results showed that the bearing with the spiral groove has better capacity performance. C.M. Rodkiewick et al. [4] conducted experiments on the influence of the groove structure on the bearing performance and concluded that the inclined groove structure can improve the bearing capacity. Yoshio et al. [5-6] compared the pressure distribution of bearings with the groove structure and smooth-surface bearings under the same working conditions. It is concluded that the comprehensive performance of bearings with the groove structure is better than that of smooth-surface bearings. In addition, V.O. Dzyura et al. [7] found that the manufacturing accuracy and surface roughness between the groove structures will also affect the carrying loading capacity and the service life of journal bearings.

In this paper, the the capacity performance of journal bearings with the spiral groove is studied. The influence of the spiral groove shape on the bearing capacity is discussed by design different spiral groove depths, widths, numbers, and pitches. This can provide technical support for the reasonable selection of the structure parameters of the spiral groove bearings.

2. Theory

2.1. Governing equations

The basic equations are used to solve all flow problems as follows:

Mass conservation equation

$$\frac{\partial \rho}{\partial t}+\nabla\cdot(\rho\vec{v})=0 \tag{1}$$

where ρ indicates the fluid density and $\vec{v}$ is the fluid velocity vector.

Momentum conservation equations

$$\frac{\partial(\rho\vec{v})}{\partial t}+\nabla\cdot(\rho\vec{v}\vec{v})=-\nabla p+\nabla\cdot(\tau)+\rho\vec{g}+\vec{F} \tag{2}$$

where p means the static pressure, τ is the stress tensor, $\rho\vec{g}$ and $\vec{F}$ are the gravitational body force and external body forces respectively. The stress tensor τ is defined as follows:

$$\tau=\mu\left[\left(\nabla\vec{v}+\nabla\vec{v}^{T}\right)-2/3\nabla\cdot\vec{v}I\right] \tag{3}$$

where μ is the fluid viscosity, I represents the unit tensor.

2.2. The load carrying capacity

The load carrying capacity can be evaluated by integrating the pressure over the journal surface, which is defined as follows:

$$F_x=-\int_0^{2\pi}\int_{-L/2}^{L/2} p\cos\varphi \mathrm{R}\mathrm{d}\varphi \mathrm{d}z \tag{4}$$

$$F_y=-\int_0^{2\pi}\int_{-L/2}^{L/2} p\sin\varphi \mathrm{R}\mathrm{d}\varphi \mathrm{d}z \tag{5}$$

$$F=\sqrt{F_x^2+F_y^2} \tag{6}$$

where F is the load carrying capacity, L is bearing length, R is the radius of the journal.

Then, the attitude angle is calculated as:

$$\theta=\arctan c\left(\frac{F_y}{F_x}\right) \tag{7}$$

3. Simulation

Due to the complex structure of the spiral groove bearing, the full fluid field simulation calculation is applied to solve the complex fluid movement by ANSYS FLUENT. The simulation model includes three parts: circular fluid field, spiral groove fluid field, and oil inlet. The top of the oil inlet is set as the fluid inlet. Two end faces of the circular fluid field and spiral groove fluid field are the fluid outlets. The inner surface of the circular fluid field is set as the rotating surface. The rotational speed is set as 16103 rpm. In the simulation model, different spiral groove structure parameters including the depth, width, number and pitch are chosen to compare the capacity performance of journal bearings. The range of the spiral groove depth is 0.2-0.4 mm. The spiral groove width is designed in the range of 2-3 mm. The spiral groove numbers are set to 1, 2, 3 respectively. Single spiral groove pitch and 0.5 times the pitch are applied in the geometry model. Then, the geometry model is meshed by the hexahedral grid. The grid number is 4732994 and the node number is 6918282. With this grid density, the fluid field calculation can achieve stable convergence. The simulation model is shown in Figure 1. The simulation parameters are listed in Table 1.

For the numerical calculation, cavitation phenomenon is modelled by Rayleigh-Plesset model for its stability and liable to convergence. For the cavitation of oil-lubricated journal bearing is mostly caused by the escape of dissolved air. The cavitation pressure of lubricating oil is set to 0 Pa. The bearing material is aluminum.

Table 1. Simulation parameters.

Parameter	Value
Bearing diameter (mm)	35
Bearing length (mm)	108
Spiral groove depth (mm)	0.2-0.4
Spiral groove width (mm)	2-3
Spiral groove number	1-3
Spiral groove pitch (mm)	45
Relative clearance	1.5‰
Rotational speed (rpm)	16103

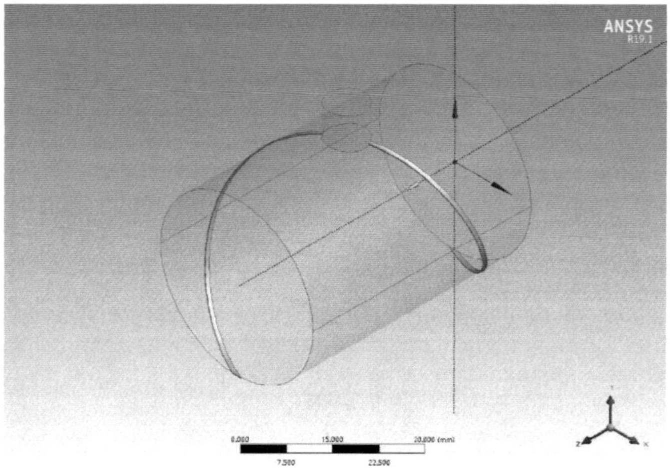

Figure 1. The simulation model

4. Results and discussion

4.1. The influence of the spiral groove depth on the bearing capacity

The influence of the spiral groove depth on the bearing capacity is studied with the minimum oil film thickness of 5 μm. The spiral groove depth is set to 0.2 mm, 0.25 mm, 0.3 mm, 0.35 mm, and 0.4 mm. As shown in Figure 2(a), the results show that with the increase of the spiral groove depth, the bearing capacity gradually decreases. The maximum carrying force decreases from 16731 N to 16532 N, which declines by 1.2%. The maximum carrying force occurs when the attitude angle is 42° or 45°. The relationship between the oil film pressure, the spiral groove depth, and the attitude angle is shown in Figure 2(b). The oil film pressure shows a uptrend when the attitude angle is in the range of 33°-51°. Besides, the oil film pressure decreases with the increase of the spiral groove depth. When the attitude angle is 51° and the spiral groove depth is 0.2 mm, the maximum oil film pressure is obtained. When the attitude angle is 33° and the spiral groove depth is 0.4 mm, the oil film pressure reaches the minimum value. Therefore, when the spiral groove depth is 0.2 mm, the bearing at the 42° attitude angle shows good capacity performance.

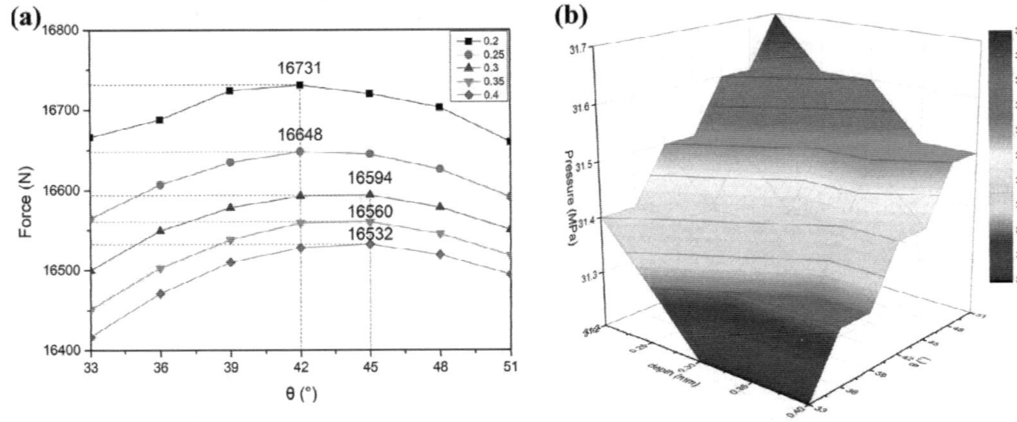

Figure 2. The influence of the spiral groove depth on the bearing capacity: (a) the bearing carrying force; (b) the oil film pressure.

4.2. The influence of the spiral groove width on the bearing capacity

Figure 3(a) shows the relationship between the attitude angle and the bearing carrying force under different spiral groove widths. The spiral groove width is set to 2 mm, 2.25 mm, 2.5 mm, 2.75 mm, and 3 mm. It can be found that the bearing capacity gradually decreases with the increase of the spiral groove width. The maximum carrying force decreases from 16532 N to 16301 N, which declines by 1.4%. The maximum carrying force occurs when the attitude angle is 42° or 45°. Figure 3(b) shows the oil film pressure in the range of 33-51° for the bearing attitude angle when the spiral groove width is 2 mm, 2.25 mm, 2.5 mm, 2.75 mm, and 3 mm respectively. The oil film pressure indicates a rising trend when the attitude angle is in the range of 33°-51°. The oil film pressure decreases with the increase of the spiral groove width. When the attitude angle is 51° and the spiral groove width is 2 mm, the oil film pressure reaches the maximum value. When the attitude angle is 33° and the spiral groove depth is 3 mm, the minimum oil film pressure is obtained. The results show that when the spiral groove width is 2 mm and the attitude angle is 45°, better bearing capacity performance is obtained.

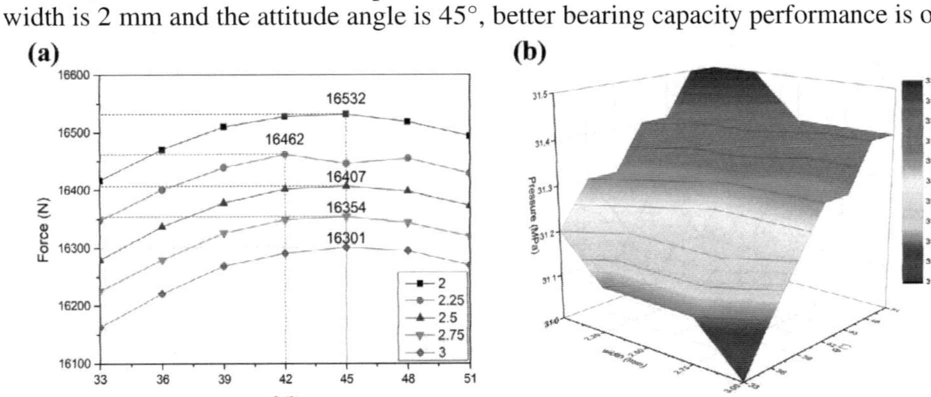

Figure 3. The influence of the spiral groove width on the bearing capacity: (a) the bearing carrying force; (b) the oil film pressure.

4.3. The influence of the spiral groove number on the bearing capacity

To study the influence of the spiral groove number on the bearing capacity performance, the spiral groove number N is set to 1, 2, 3. Figure 4(a) shows that when the spiral groove number increases from 1 to 2 to 3, the corresponding maximum carrying force decreases from 16478 N to 8934 N to 5133 N. The maximum carrying force of double spiral groove is decreased by 45.8% compared with that of single spiral groove. While the maximum carrying force of three spiral groove is decreased by 68.8% compared with that of single spiral groove. As shown in Figure 4(b), when the spiral groove number is 1, 2, 3, the maximum oil film pressure is 31.4MPa, 21.3MPa, 13.7 MPa respectively. The maximum oil film pressure of double spiral grooves is decreased by 32.2% compared with that of single spiral groove. While the maximum oil film pressure of three spiral grooves is decreased by 56.4% compared with that of single spiral groove. Therefore, the bearing capacity can be improved by decreasing the spiral groove number. The oil film pressure distribution under different spiral groove numbers is shown in Figure 5. It can be investigated that after the fluid field is divided by the spiral grooves, the oil film pressure in each bearing carrying load region is evenly distributed, and the low pressure region is fully expanded.

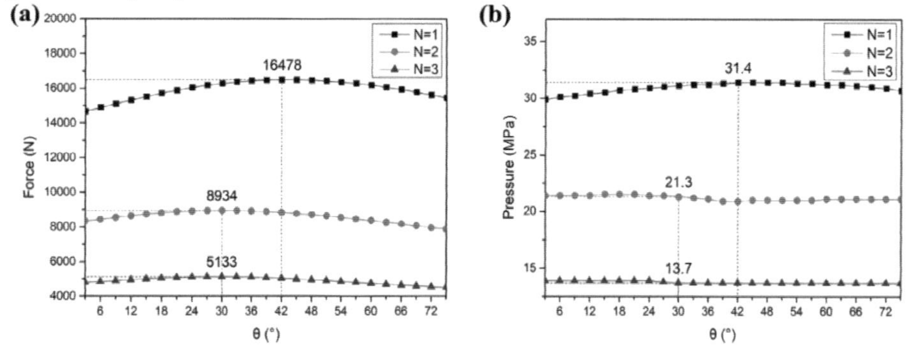

Figure 4. The influence of the spiral groove number on the bearing capacity: (a) the bearing carrying force; (b) the oil film pressure.

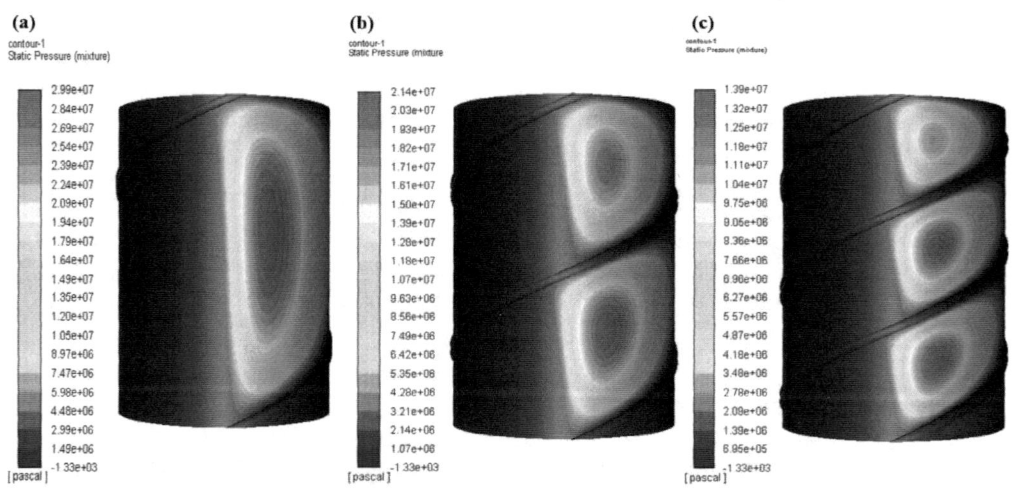

Figure 5. The oil film pressure distribution under different spiral groove numbers: (a) single spiral groove; (b) double spiral grooves; (c) three spiral grooves.

4.4. The influence of the spiral groove pitch on the bearing capacity

Three spiral groove bearing structures are established to discuss the influence of the spiral groove pitch on the bearing capacity performance. Three bearing structures are single spiral groove with single pitch, double spiral groove with single pitch, and single spiral groove with 0.5 times the pitch. Figure 6(a) shows the relationship between the carrying force and the offset angle under different spiral groove pitches. It can be observed that when the spiral groove number is same, the carrying force decreases with the decrease of the pitch. The single spiral groove structure with single pitch at 42° attitude angle has the maximum carrying force of 16478N. While the double spiral groove structure with single pitch at 30° attitude angle has the maximum carrying force of 8934 N. When the pitch is decreased by 0.5 times, the carrying force of single spiral groove bearing is 10777 N, which is decreased by 34.6% compared with that of the single pitch. As shown in Figure 6(b), the maximum oil film pressure of 31.4MPa is obtained when the bearing structure is the single spiral groove with single pitch. The maximum oil film pressure of the double spiral groove with single pitch is 21.3MPa. The maximum oil film pressure of the single spiral groove with 0.5 times the pitch is 22.9MPa. Compared with the single spiral groove with single pitch, the maximum oil film pressure of the double spiral groove with single pitch and the single spiral groove with 0.5 times the pitch is declined by 32.2% and

27.1%, respectively.

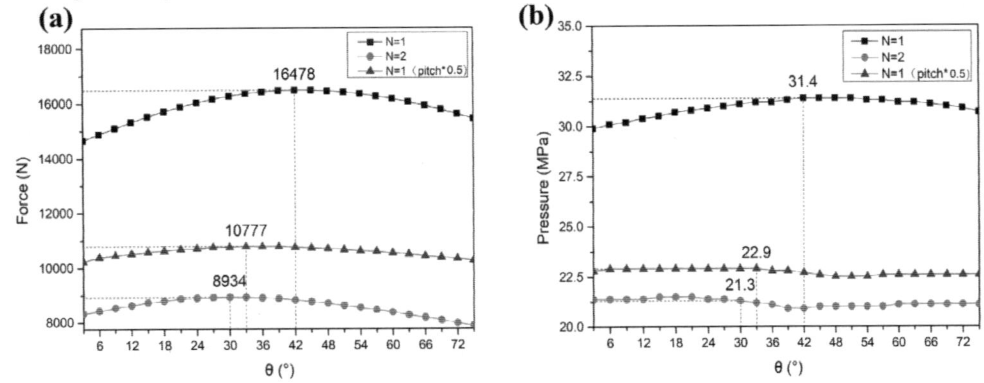

Figure 6. The influence of the spiral groove pitch on the bearing capacity:
(a) the bearing carrying force; (b) the oil film pressure.

The oil film pressure distribution under different spiral groove pitches is shown in Figure 7. The spiral groove divides the high oil film pressure zone obviously. Because different pitches correspond to different spiral angles, therefore the spread form of the high oil film pressure zone is significantly affected by the spiral angles of the spiral grooves. For the single spiral groove with 0.5 times the pitch, the area of the high oil film pressure zone is larger than that of the single spiral groove with single pitch.

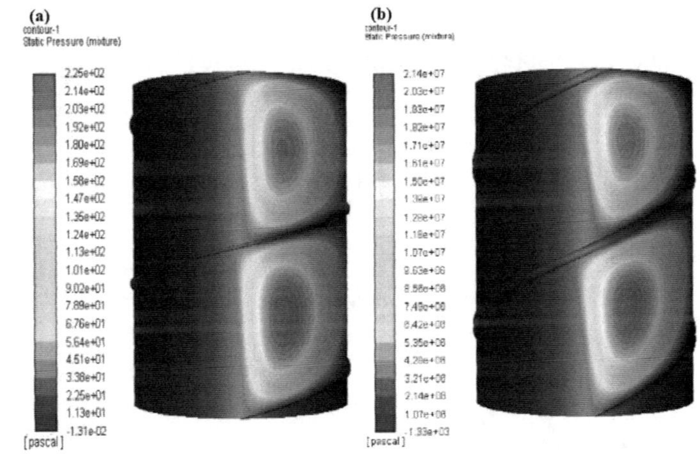

Figure 7. The oil film pressure distribution under different spiral groove pitches:
(a) single spiral groove with 0.5 times the pitch; (b) single spiral groove with single pitch.

5. Conclusion

In this paper, the influence of the spiral groove bearing structure on the bearing capacity performance is discussed. The main conclusions are as follows:

(1) With the increase of the spiral groove depth, the bearing capacity gradually decreases. When the spiral groove depth is 0.2 mm, the bearing at the 42° attitude angle shows good capacity performance. Besides, with the increase of the spiral groove width, the bearing capacity gradually decreases. When the spiral groove width is 2 mm and the attitude angle is 45°, better bearing capacity performance is obtained.

(2) As the spiral groove numbers increase, the maximum carrying force and maximum oil film pressure decreases and the bearing capacity becomes worse. The oil film pressure distribution is obviously divided by the spiral grooves. The area of the high oil film pressure zone shrinks significantly.

(3) As the spiral groove pitches increase, the maximum carrying force and maximum oil film pressure increases, which shows good bearing capacity performance. The area of the high oil film pressure zone is also expanded.

Acknowledgement
The authors acknowledge the financial support provided by Marine Low Speed Engine Project-Phase I (Grant No. CDGC01-KT11).

References

[1] SINGH, Narendra, SHARMA, Performance of membrane compensated multirecess hydrostatic/hybrid flexible journal bearing system considering various recess shapes, Tribol. Int. 37 (1) (2004) 11-24.

[2] Sharma S C, Sinhasan R, Bharuka D K, Performance of Hydrostatic/Hybrid Journal Bearings with Unconventional Recess Geometries, Tribology Transaction, 41 (3) (1998) 375-381.

[3] Sharma S C, Jain S C, Bharuka, Influence of recess shape on the performance of a capillary compensated circular thrust pad hydrostatic bearing, Tribol. Int. 35 (2007) 347-356.

[4] Rodkiewicz C M, Dalita W. Experimental Investigation Regarding the Effects of Grooves on Conical Bearing Performance, Tribology Transactions, 138 (1) (1995) 178-182.

[5] Kumada Y, Hashzume K, Kimura Y. Performance of Plain Bearings with Circumferential Microgrooves, TLE Tribology Transactions, 1 (39) (1995) 81-86.

[6] Watanabe K, Natsume J, Hashizume K, Tjepretocal analysis of bearing performance of microgrooved bearing, JASE Review, 21 (2000) 29-34.

[7] V. Dzyura, P. Maruschak, I. Zakiev, A. Sorochak. Analysis of inner surface roughness parameters of load-carrying and support elements of mechanical systems, International Journal of Engineering, Transactions B: Applications, 30 (8) (2017) 1170-1175.

Design of a separation machine using pneumatic system combined with sieve vibration for removing parchment coffee from Robusta green coffee bean

W Limmun and T Chungcharoen*

Department of Engineering, King Mongkut's Institute of Technology Ladkrabang, Prince of Chumphon Campus, 17/1, Chumko Pathio, Chumphon, 86160, Thailand

*Corresponding Author, E-mail: thatchapol.ch@kmitl.ac.th

Abstract. The hulling process for coffee bean extraction may not effectively release the bean from parchment. This might negatively affect the aroma or taste of coffee in the coffee beans roasting process. Consequently, the objectives of this work were to design and construct a machine for separating parchment coffee from Robusta green coffee bean using a pneumatic separation system combined with sieve vibration. The performance of the separation machine was investigated under different operating conditions such as air velocity, vibration frequency, and the angle of sieve inclination. The samples were fed into the machine at a feed rate of 20 kg/h. The testing conditions were set at air velocities of 2 to 4 m/s, vibration frequencies of 36 to 40 Hz, and sieve inclination angle of 4° to 8°. The results showed that the purity and separation efficiency of parchment increased with increasing air velocity. Conversely, decreasing vibration frequency and sieve angle increased the purity and separation efficiency. The appropriate conditions for parchment coffee separation from Robusta green coffee bean using a pneumatic-sieve vibration system were 4 m/s air velocity, 36 Hz vibration frequency, and 8° sieve angle with the highest overall effectiveness of 92%.

1. Introduction

Coffee has major economic significance for both production and consumption in the Thailand markets. Robusta coffee or Coffea canephora reports a high production of coffee in Thailand [1]. There is major export production in southern Thailand, which amounted to about 25 thousand bags in February 2020 [2], mainly used as a raw material of instant coffee. The coffee cherries go through the coffee bean processing into the finished coffee ready for the roasting process by several methods such as dry method, semi-dry method, and wet method [3]. However, the dry method is widely used because of its simplicity, less water, and machinery requirements. The dry method has two steps; drying and hulling the cherries to produce green beans. The hulling process properly removes cherry fruit (skin, pulp, and parchment) from the bean via hand mills or hulling machines. The removing parchment coffee from Robusta green coffee beans can have a significant effect on the final quality of the green coffee since it is playing an important role in aroma formation or taste during the roasting process. After hulling, coffees are sorted into screen sizes, and defect sorting takes place.

In the hulling process, the sorting of parchment coffee from the green coffee bean is done manually, using vibrating sieves, and pneumatic separation system to separate the parchment coffee from green coffee beans before milling. The vibrating sieves operation is a screening process based on the movement of materials, the mesh size of the sieve, and the direction of movement of the sieve [4]. However, there might result in increased grain losses since the grains stick together on the sieve [5, 6].

The pneumatic separation involves lifting up lightweight materials from the grain, while heavier materials move downward by the airflow [7]. However, the low degree of separation is a limitation of the pneumatic separation system [8]. Additionally, the pneumatic separation system is currently ineffective and unsatisfactory since green coffee beans are small, resulting in low porosity values. High separation efficiency for the pneumatic system is probably due to the installation of specially designed devices that stabilize the uniformity of coffee bean distribution. Therefore, a vibrating separator with a sieve is a good alternative to distribute coffee during the hulling process.

Generally, the hulling process may not effectively release the bean from parchment. Therefore, this study aimed to design and construct a machine for separating parchment coffee from Robusta green coffee bean using a pneumatic separation system combined with sieve vibration. The performance (purity of separated parchment, separation efficiency, and overall effectiveness) of the separation machine was investigated under different operating conditions such as air velocity, vibration frequency, and sieve angle.

2. Materials and methods

2.1. Materials

Robusta green coffee beans and parchment coffee were produced and harvested in Chumphon Province in the south of Thailand, as shown in figure 1. Robusta green coffee beans were used in this study without the size selection process and a moisture content of 9.8% dry basis. The moisture content of parchment coffee was 8.5% dry basis. The samples, 0.9 kg green coffee bean, and 0.1 kg parchment coffee, were used, and the green coffee beans in the samples were sprayed to clearly identified samples.

Figure 1. Robusta green coffee beans without a parchment coffee (a), parchment coffee (b), and the samples (a mixture of green coffee beans and parchment coffee) (c)

2.2. Experimental apparatus

The separation of the parchment coffee and Robusta green coffee bean is done on sifting sieves supported with a pneumatic system, as shown in figure 2. The design of the machine involved developing and assessing the performance of separation. The separation machine consists of a pneumatic system of rectangular cross-section width of 300 mm and a depth of 150 mm. A pneumatic system is supplied by a centrifugal blower with 1 hp, which is s fitted below the sieve vibration system. The sieve vibration system is 90 cm in length and 70.2 cm in width with an aperture size of 0.3 mm. The sieve was installed on an inclined plate and connected to a frame by bolts to change the angle of sieve inclination (2). Two motors drive the vibration system with 0.18 kW.

The samples (a mixture of green coffee beans and parchment coffee) were fed into the hopper (1) to be distributed and separated by a pneumatic system combined with sieve vibration. The sieve vibration system and sieve inclination angle were adjusted, then the experiment was started. All experiments were conducted with the samples at a feed rate of 20 kg/h. Afterwards, green coffee bean and parchment coffee were collected separately in outlets (3) and (4) and weighed using electronic balances.

IOP Conf. Series: Materials Science and Engineering **999** (2020) 012016 doi:10.1088/1757-899X/999/1/012016

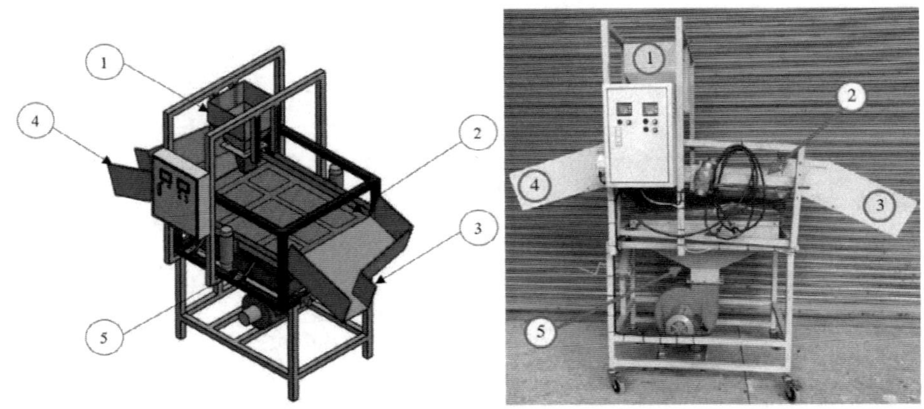

Figure 2. Isometric view of separation machine using pneumatic system combined with sieve vibration: 1-hopper, 2-sieving vibration unit, 3-green coffee bean outlet, 4-parchment coffee outlet, 5-pneumatic system.

2.3. Experimental procedure

The performance of the separation machine was investigated under different operating conditions such as air velocity, vibration frequency, and sieve angle. The samples were fed into the machine at a feed rate of 20 kg/h. The testing conditions were set at air velocities of 2 to 4 m/s, vibration frequencies of 36 to 40 Hz, and sieve inclination angle of 4° to 8°. The number of green coffee bean and parchment coffee was recorded to evaluate the performance of the separation machine. The purity of separation, separation efficiency, and overall effectiveness was determined using equations (1)–(11).

The purity of separated coffee bean or parchment coffee was determined by the ratio of the coffee bean or parchment coffee to the total weight of their samples, which fed into the hopper using the formula below:

$$P_{CB} = \left(\frac{W_{CB} - Q_{PC}}{W_{CB}}\right) \times 100 \tag{1}$$

$$P_{PC} = \left(\frac{W_{PC} - q_{CB}}{W_{PC}}\right) \times 100 \tag{2}$$

where: P_{CB} – the percentage purity of coffee bean after separation, %;
P_{PC} – the percentage purity of parchment coffee after separation, %;
W_{CB} – the weight of all samples in the coffee bean outlet, g;
W_{PC} –the weight of all samples in the parchment coffee outlet, g;
q_{CB} – the weight of coffee bean outlet in the parchment coffee outlet, g;
q_{PC} – the weight of parchment coffee in coffee bean outlet, g;

The separation efficiencies of the coffee bean (ECB) and parchment coffee (EPC) were determined based on the weight of mixture components. The separation efficiency of each component was calculated with the formula below:

$$E_{CB} = \frac{W_{CB} - q_{PC}}{Q_{CB}} \times 100 \tag{3}$$

$$E_{PC} = \frac{W_{PC} - q_{CB}}{Q_{PC}} \times 100 \tag{4}$$

where: Q_{CB} – the weight of coffee bean, g;
Q_{PC} – the weight of parchment coffee, g.

The overall effectiveness (η) is a measure of performance of the separating parchment coffee from a coffee bean, indicating a relationship between the separating ability, purity, and separation efficiency. The overall effectiveness was calculated with the formula below:

$$\eta = \left[F_{CB} \frac{(P_{CB}\text{-}a_{CB})}{1\text{-}a_{CB}} + F_{PC} \frac{(P_{PC}\text{-}a_{PC})}{1\text{-}a_{PC}} \right] \times 100 \tag{5}$$

The separating ability of material, purity, and separation efficiency was determined with the following formula:

$$F_{CB} = \frac{W_{CB}}{W_T} \tag{6}$$

$$F_{PC} = \frac{W_{PC}}{W_T} \tag{7}$$

where: F_{CB} – separating ability of material in coffee bean outlet;
F_{PC} – separating ability of material in parchment coffee outlet;
W_T – the weight of all samples, g.

$$p_{CB} = \frac{W_{CB}\text{-}q_{PC}}{W_{CB}} \tag{8}$$

$$p_{PC} = \frac{W_{PC}\text{-}q_{CB}}{W_{PC}} \tag{9}$$

where: p_{CB} – purity of coffee bean after separation, %;
p_{PC} – purity of parchment coffee after separation, %.

$$a_{CB} = \frac{Q_{CB}}{W_T} \tag{10}$$

$$a_{PC} = \frac{Q_{PC}}{W_T} \tag{11}$$

where: a_{CB} – the ratio of coffee bean weight to the total weight of the samples;
a_{PC} – the ratio of parchment coffee weight to the total weight of the samples.

All experiments were carried out triplicate, and all data were expressed as the mean ± SD. The differences in mean values were evaluated by one-way analysis of variance (ANOVA) using SPSS. The result of ANOVA does not provide detailed information regarding the differences among various combinations of groups. The analysis of the differences between particular pairs of experimental groups was performed by using the LSD method as the multiple comparison test. The values with the same letter were significantly different ($P < 0.05$).

3. Results and discussion

3.1. The separation efficiency of green coffee bean and parchment coffee

The influence of various operating parameters were investigated at air velocities of 2 to 4 m/s, vibration frequencies of 36 to 40 Hz, and sieve inclination angle of 4° to 8°. The variations in the separation efficiencies of green coffee bean and parchment coffee are presented in figures 3 and 4. The separation efficiencies of the green coffee bean at a sieve inclination angle of 4° and vibration frequency of 36 Hz decreased from 76.1% to 65.7% when the air velocity increased from 2 to 4 m/s. However, the separation efficiencies of the parchment coffee at a sieve inclination angle of 4° and vibration frequency of 36 Hz increased from 87.4% to 96.7% when the air velocity increased from 2 to 4 m/s. The separation efficiency of the green coffee bean slightly decreased with increasing air velocity, while the separation efficiency of parchment coffee increased. These results indicated that the green coffee bean and parchment coffee might be passing through the parchment coffee outlet at high air velocity in a pneumatic system.

On the other hand, the separation efficiency of the green coffee bean at a sieve inclination angle of 8° increased to 99.2% at a high air velocity of 4 m/s and vibration frequency of 40 Hz. In contrast, the separation efficiency of parchment coffee decreased to 89.3%. This suggested that the green coffee bean and parchment coffee had been quickly distributed by sieve vibration, which restricted the pneumatic system from blowing the coffee beans into the wrong outlet channel. In addition, the increase in air velocity coupled with an increasing sieve inclination angle from 4° to 8° might have caused the rising movement of coffee beans into the coffee bean outlet. Our result is consistent with Srisang et al. [9] that the sieve inclination angle significant effect on separation efficiency. As a result, the separation efficiency of the green coffee bean was found to have increased with increasing the angle of inclination.

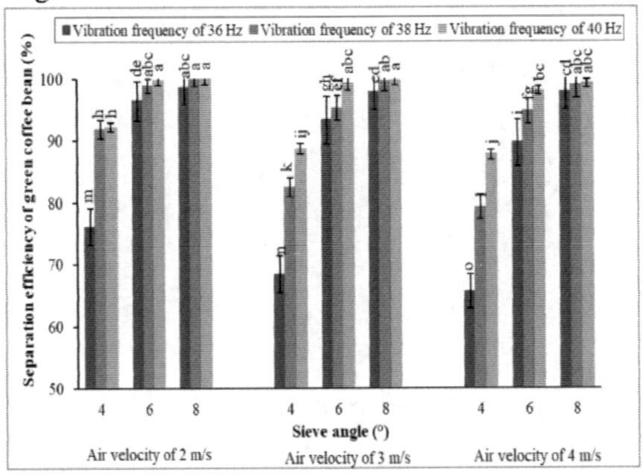

Figure 3. Separation efficiency of green coffee bean with different sieve inclination angles, air velocities, and vibration frequencies.

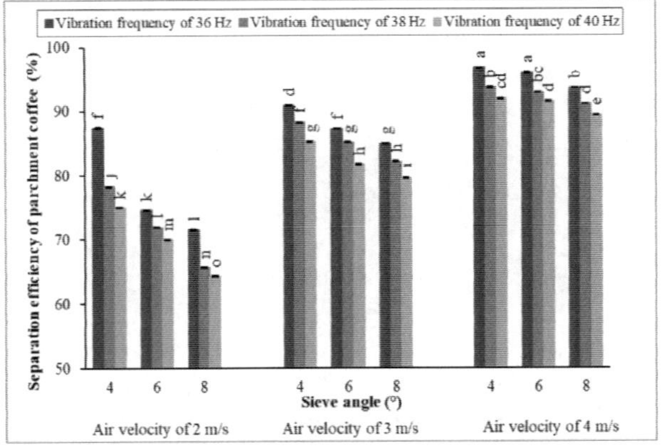

Figure 4. Separation efficiency of parchment coffee with different sieve inclination angles, air velocities, and vibration frequencies.

3.2. The purity separation of green coffee bean and parchment coffee

The amount of green coffee bean and parchment coffee in both outlets were weighed to assess the purity of separated parchment coffee from green coffee beans. The percentage purity of separation presented in figures 5 and 6 at different sieve inclination angle, and vibration frequency, the purity separation of green coffee bean increased with increasing air viscosity from 2 to 4 m/s. These suggested that high air velocity is more efficient in separating parchment coffee from coffee beans. However, the parchment coffee outlet still had a mixing of samples due to high air velocity conditions. This indicated that the purity of separated parchment coffee slightly decreased (figure 6). The purity separation of green coffee beans gradually decreased when vibration frequency increased from 36 to 40 Hz and at different levels of sieve inclination angle since the parchment coffee was also found in the green coffee bean outlet. These results were consistent with the significant drop in the separation efficiency of parchment coffee in the same conditions. However, the purity separation of parchment coffee increased when the vibration frequency and sieve inclination angle increased.

3.3. The overall effectiveness of green coffee bean and parchment coffee

The overall effectiveness of the separation was calculated from the separation efficiency and purity of separated of coffee bean and parchment coffee. The variations of the overall effectiveness of removing parchment coffee from Robusta green coffee bean under different conditions are shown in figure 7. The overall effectiveness of this separation machine using a pneumatic system combined with sieve vibration ranged between 59.3 and 92.2%. The appropriate condition for parchment coffee separation from green coffee bean was 4 m/s air velocity, 36 Hz vibration frequency, and 8° sieve inclination angle with the highest overall effectiveness of 92.2%. In this condition, the separation efficiencies of the green coffee bean and parchment coffee were 97.8% and 93.6%, respectively, and the purities separation of the green coffee bean and parchment coffee were 99.4% and 82.6%, respectively.

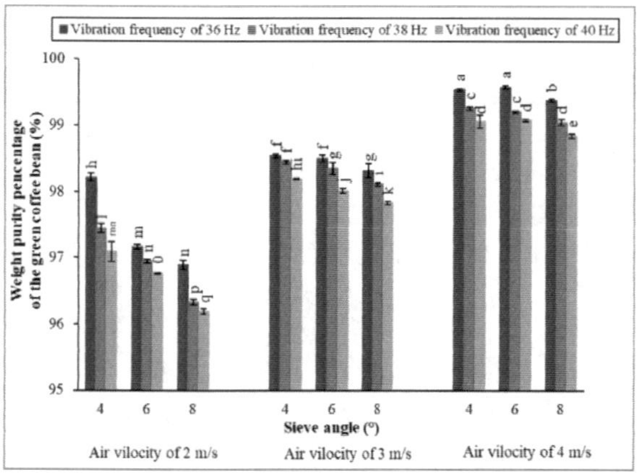

Figure 5. Percentage purity of separated green coffee bean with different sieve inclination angles, air velocities, and vibration frequencies.

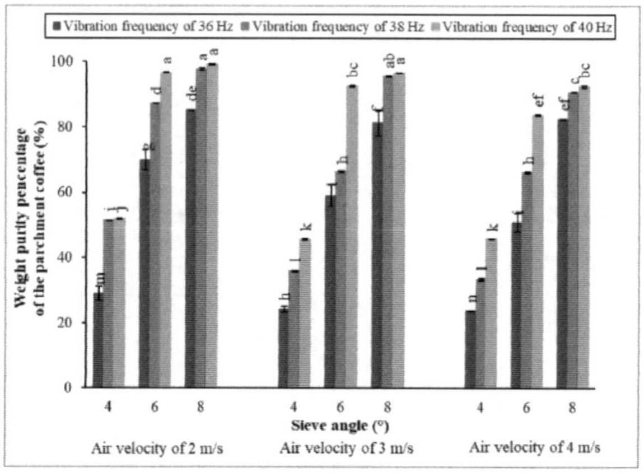

Figure 6. Percentage purity of separated parchment coffee with different sieve inclination angles, air velocities, and vibration frequencies.

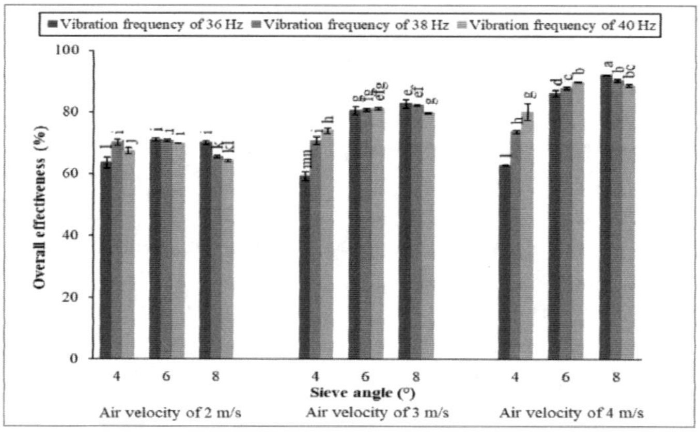

Figure 7. Overall effectiveness of removing parchment coffee from Robusta green coffee bean with different sieve inclination angles, air velocities, and vibration frequencies.

4. Conclusions

This study was designed to investigate parameters affecting the separation efficiencies for the novel separation machine for removing parchment coffee from Robusta green coffee beans. The operating conditions such as air velocities of 2 to 4 m/s, vibration frequencies of 36 to 40 Hz, and sieve inclination angle of 4° to 8° were conducted using a separation machine with a pneumatic system combined with sieve vibration. The results show that increasing air velocity and decreasing vibration frequency and sieve inclination angle in separating parchment coffee from Robusta green coffee beans increased the separation efficiency of parchment coffee and the purity separation of green coffee beans. The highest overall effectiveness for parchment coffee separation from green coffee bean was 92.2% at an air velocity of 4 m/s, vibration frequency of 36 Hz, and sieve inclination angle of 8°. Additionally, further work on the size grading of green coffee beans using these methods is needed to find optimal operating conditions that enhance the performance of coffee processing and designing of the separating machine.

Acknowledgments

The authors would like to acknowledge King Mongkut's Institute of Technology Ladkrabang, Prince of Chumphon Campus.

IOP Conf. Series: Materials Science and Engineering **999** (2020) 012016 doi:10.1088/1757-899X/999/1/012016

References

[1] Nitthikan N, Leelapornpisid P, Natakankitkul S, Chaiyana W, Mueller M, Viernstein H and Kiattisin K 2018 *J. Nanotechnol.* **2018**.
[2] International Coffee Organization 2020 *Trade Statistics Tables, Monthly export statistics - February 2020* (London: ICO)
[3] Giselle S. D, Antônio A. P and Adriana F 2010 *Food Chem.* **118** 851
[4] Chanpaka W, Srisang N, Dangwilailux P and Chungcharoen T 2020 *J. Phys. Conf. Ser.* **1519** 012011
[5] Lawinska K and Modrzewski R 2017 *Physicochem. Probl. Miner. Process.* **53(2)** 812
[6] Piven V. V 2019 *Eurasian J. Biosci.* **13** 1391
[7] Innocentini M. D. M; Barizan W. S., Alves M. N. O, Jr R. P 2009 *FOOD BIOPROD PROCESS* **87** 237
[8] Kroulík M, Hůla J, Rybka A and Honzík I 2016 *Res. Agr. Eng.* **62** 56
[9] Srisang N,Chanpaka W and Chungcharoen T 2019 *IOP Conf. Ser.: Earth Environ. Sci.* **301** 012037

IOP Conf. Series: Materials Science and Engineering **999** (2020) 012017 doi:10.1088/1757-899X/999/1/012017

Buckling of micropolar beams by an improved first order deformation theory

K N Betancourt[a], K Soncco[b] and R Arciniega[1,c]

[1] Department of Civil Engineering, Universidad Peruana de Ciencias Aplicadas, Lima, Peru
Email: u201410150@upc.edu.pe[a]; u201412519@upc.edu.pe[b]; roman.arciniega@upc.edu.pe[c]

Abstract. In this paper, we present a variational formulation to study the buckling behavior of micropolar beams by using an improved 3D deformation theory. A micropolar continuum applied to beams has been developed using its natural Lagrangian kinematic relations. The Rodriguez rotation measure was used to describe the rotational degrees of freedom. Additionally, a Taylor expansion was performed to linearize the kinematic relations. For the buckling analysis, the Trefftz criterion procedure was applied. A finite element model was derived for the solution of the variational problem using spectral interpolation functions for a higher convergence rate and for avoiding shear locking problems. The results describe the influence of the micropolar parameters and size-dependent behavior. Finally, the model was used to evaluate the buckling loads of simply-supported functionally graded beams considering experimental material parameters.

1. Introduction

Nowadays, the discoveries and applications of new materials that evidence local and non-local mechanical behaviors have increased. Moreover, as new technologies are developed, the use of nanocomponents and nanostructures is preferred. As size decreases, several studies have shown the importance and influence of size-dependent mechanical models [1,2]. A Cosserat continuum [3] is related to a non-classical scheme because it uses independent rotations that allow the evaluation of microstructures and their size-dependent behavior. Applications of micropolar or Cosserat continuum have been developed in civil engineering [4,5], soil mechanics [6,7], and more.

Studies regarding micropolar mechanics have mainly focused on bending and vibration responses. Ansari et al. [8] applied the Finite Element Method for the analysis of microbeams free vibration response. They proposed a nonclassical 3D element for studying the influence of micropolar parameters on vibration frequencies. Penta [9] analyzed the buckling behavior of Euler Bernoulli Vierendeel beams using a homogenized theory. His study has shown the influence of the micropolar continuum on the critical load of Vierendeel girder beams. Carrera and Zozulya [10] developed a unified beam theory for micropolar schemes on analytical bending responses. Ramezani et al. [11] studied the bending response of cantilever micropolar beams. The present authors [12] also investigated the bending response of micropolar beams, comparing their improved shear deformation theory against Hassanpour's [13] formulation.

Functionally graded materials (FGM) were first discovered in Japan [14], which are the result of the gradual variation of two phases along the desired direction. These materials enhance mechanical and thermal properties and have been applied in aerospace engineering, tribology, nanotechnology, and biology. Even though there are several studies about FGM beams, there are few on micropolar buckling models.

IOP Conf. Series: Materials Science and Engineering **999** (2020) 012017 doi:10.1088/1757-899X/999/1/012017

The current research proposes a model to perform buckling analysis using the incremental-fundamental procedure (Trefftz criterion) on micropolar beams through a variational formulation. Several examples are provided and verified with the literature. The authors show numerical results for FGM beams to evaluate the influence of the power-law index in the buckling load, by using the following mixture rule and power-law function through the thickness x^2 of the beam

$$P(x^2) = P_b V_b + P_t V_t, \qquad V_t = (x^2/h + 1/2)^n, \qquad V_b = 1 - V_t. \tag{1}$$

where P_b and P_t are the material properties at the bottom and top layers of the element, V_b and V_t are their respective volume fractions and n is the power-law index.

2. Theoretical Formulation

2.1. Micropolar theory

In the micropolar theory, the kinematics relations of a point are characterized by rotational and transactional degrees of freedom. The nonlinear kinematic relations can be found on Eringen's work [15], in which he defines a stretch and wryness tensor as:

$$\mathbf{E} = \mathbf{F}^{\mathrm{T}}\mathbf{Q},$$
$$\boldsymbol{\Gamma} = -\frac{1}{2}\epsilon : \left(\mathbf{Q}^{\mathrm{T}} Grad\mathbf{Q}\right), \tag{2}$$

where $\mathbf{F}$ is the deformation gradient, $\mathbf{Q}$ is the third-order proper orthogonal microrotation tensor, and ϵ is the third-order skew Ricci tensor.

2.2. Micropolar material parameters

For a linear micropolar isotropic solid, stress-strain relations for the stress tensor $\boldsymbol{\sigma}$ and the couple stress tensor $\boldsymbol{\chi}$ are stated as [16]

$$\boldsymbol{\sigma} = \lambda \mathbf{I}\mathrm{tr}\boldsymbol{\varepsilon} + (\bar{\mu} + \kappa)\boldsymbol{\varepsilon} + \bar{\mu}\boldsymbol{\varepsilon}^{T},$$
$$\boldsymbol{\chi} = \beta_1 \mathbf{I}\mathrm{tr}\boldsymbol{\gamma} + \beta_2 \boldsymbol{\gamma}^{T} + \beta_3 \boldsymbol{\gamma}, \tag{3}$$

where $\bar{\mu}, \lambda, \kappa, \beta_1, \beta_2, \beta_3$ micropolar elastic constants. Parameters $\bar{\mu}$ and κ relates to the Lamé shear modulus μ according to [17] as

$$\mu = \bar{\mu} + \frac{1}{2}\kappa, \tag{4}$$

and λ is the first Lamé parameter. Furthermore, Young's modulus E and Poisson's ratio v can be defined in terms of these micropolar elastic constants by the following equations

$$E = \frac{(2\bar{\mu} + \kappa)(3\lambda + 2\bar{\mu} + \kappa)}{2\lambda + 2\bar{\mu} + \kappa}, \nu = \frac{\lambda}{2\lambda + 2\bar{\mu} + \kappa}. \tag{5}$$

2.3. Beam theory

Let $\{x^i\}$ be a set of Cartesian coordinates with orthonormal basis $\{\mathbf{e}_i\}$. x^1 is the neutral axis of the beam and x^2 is the thickness axis. The displacement and microrotation fields are assumed to be of the following form (see [18])

$$\mathbf{v}(x^1, x^2) = \mathbf{u}(x^1) + x^2\boldsymbol{\varphi}(x^1) + (x^2)^2\boldsymbol{\psi}(x^1), \tag{6}$$

IOP Conf. Series: Materials Science and Engineering **999** (2020) 012017 doi:10.1088/1757-899X/999/1/012017

$$\boldsymbol{\vartheta}(x^1) = \boldsymbol{\theta}(x^1). \tag{7}$$

where $\mathbf{u} = u_i\mathbf{e}_i$ represents the displacement vector of the neutral axis, while $\boldsymbol{\varphi} = \varphi_i\mathbf{e}_i$ and $\boldsymbol{\psi} = \psi_2\mathbf{e}_2$ are difference vectors with $i = 1,2$. The displacement field contains five independent variables and one quadratic term $\boldsymbol{\psi}$ to avoid poisson locking. The microrotation field is determined as $\boldsymbol{\theta} = \theta_3\mathbf{e}_3$, which represent the microrotation along the transversal axis x^3 of the beam.
For the given displacement and microrotation fields in equations (6) and (7) and by (2), the stretch and wryness tensors are defined according to the assumptions proposed by Hasanyan and Waas [19]. Nonlinear terms are only considered in the first component of the stretch tensor according to the following equations

$$\begin{array}{ll} \varepsilon_{11}^{(0)} = u_{1,1} + 1/2{\theta_3}^2 + \theta_3 u_{2,1} & \varepsilon_{11}^{(1)} = \varphi_{1,1} \\ \varepsilon_{22}^{(0)} = \varphi_2 & \varepsilon_{22}^{(1)} = 2\psi_2 \\ \varepsilon_{12}^{(0)} = u_{2,1} - \theta_3 & \varepsilon_{12}^{(1)} = \varphi_{2,1} \\ \varepsilon_{21}^{(0)} = \varphi_1 + \theta_3 & \varepsilon_{21}^{(1)} = 0 \\ \gamma_{13}^{(0)} = \theta_{3,1} & \gamma_{13}^{(1)} = 0 \end{array} \tag{8}$$

2.4. Principle of virtual work
The weak form of the model is constructed by the Principle of Virtual Work in which virtual stress measures are work-conjugate to virtual strain measures in the micropolar continuum, as stated in [20]. The configuration solution of the micropolar beam is defined by the set $\Phi \equiv (\mathbf{u}, \boldsymbol{\varphi}, \boldsymbol{\psi}, \boldsymbol{\theta})$. Thus,

$$\mathcal{G}(\Phi, \delta\Phi) = \mathcal{G}_{int}(\Phi, \delta\Phi) - \mathcal{G}_{ext}(\Phi, \delta\Phi),$$

$$\mathcal{G}(\Phi, \delta\Phi) = \int_{x^1} \left(\mathbf{N}^{(0)} \cdot \delta\boldsymbol{\varepsilon}^{(0)} + \mathbf{N}^{(1)} \cdot \delta\boldsymbol{\varepsilon}^{(1)} + \mathbf{M}^{(0)} \cdot \delta\boldsymbol{\gamma}^{(0)}\right) dx^1 + \int_{x^1} (\mathbf{p} \cdot \delta\mathbf{u} + \mathbf{m} \cdot \delta\boldsymbol{\theta}) dx^1, \tag{9}$$

where $\delta\Phi \equiv (\delta\mathbf{u}, \delta\boldsymbol{\varphi}, \delta\boldsymbol{\psi}, \delta\boldsymbol{\theta})$. $\mathbf{N}^{(i)}$ are the force stress resultants, $\mathbf{M}^{(0)}$ is the micropolar stress resultant, $\mathbf{p}$ are the body forces acting on the beam per unit length, and $\mathbf{m}$ are the body moment forces, respectively. The force stress resultants $\mathbf{N}^{(i)}$are of the form

$$\mathbf{N}^{(i)} = \sum_{j=1}^{1} \mathbb{B}^{(i+j)} \boldsymbol{\varepsilon}^{(j)}, \qquad i = 0,1, \tag{10}$$

while moment stress resultants $\mathbf{M}^{(0)}$ have the form

$$\mathbf{M}^{(i)} = \mathbb{M}^{(0)} \boldsymbol{\gamma}^{(0)}. \tag{11}$$

For the definition of $\mathbf{N}^{(i)}$ and $\mathbf{M}^{(0)}$, it is necessary to define the material stiffness coefficients $\mathbb{B}$ and $\mathbb{M}$ as

$$\mathbb{B}^{(k)} = \int_{-h/2}^{h/2} (x^2)^k \mathbb{C} dx^2, \qquad k = 0,1,2, \qquad \mathbb{M}^{(0)} = h\mathbb{D}, \tag{12}$$

where h represents the thickness of the beam.
The fourth order micropolar tensors $\mathbb{C}$ and $\mathbb{D}$ are stated by means of the definitions from [16]

$$\mathbb{C} = \lambda \mathbf{I}\otimes\mathbf{I} + (\bar{\mu} + \kappa)\mathbf{e}_a\otimes\mathbf{e}_b\otimes\mathbf{e}_a\otimes\mathbf{e}_b + \bar{\mu}\mathbf{e}_a\otimes\mathbf{I}\otimes\mathbf{e}_a,$$
$$\mathbb{D} = \beta_1\mathbf{I}\otimes\mathbf{I} + \beta_2\mathbf{e}_a\otimes\mathbf{I}\otimes\mathbf{e}_a + \beta_3\mathbf{e}_a\otimes\mathbf{e}_b\otimes\mathbf{e}_a\otimes\mathbf{e}_b, \qquad (13)$$

where $a, b = 1,2,3$. From equations (11) to (13), the final expression of the virtual work can be stated as

$$\mathcal{G}(\Phi, \delta\Phi) = \int_{x^1} \sum_{i=0}^{1}\sum_{j=0}^{1} \delta\boldsymbol{\varepsilon}^{(i)}\mathbb{B}^{(i+j)}\,\boldsymbol{\varepsilon}^{(j)}\,dx^1 + \int_{x^1} \delta\boldsymbol{\gamma}^{(0)}\mathbb{M}^{(0)}\,\boldsymbol{\gamma}^{(0)}dx^1 - \int_{x^1} (\mathbf{p}\cdot\delta\mathbf{u} + \mathbf{m}\cdot\delta\boldsymbol{\theta})dx^1. \qquad (14)$$

2.5. Stability analysis

The stability analysis considers two systems: the incremental and the fundamental. The former defines the perturbation and the second the membrane state [21,22]. Based on the principle of virtual work, the displacement field is defined as: $\Phi = \Phi^{\mathrm{F}} + \Phi^{\mathrm{I}}$, where F stands for fundamental and I, for incremental. For the independent displacement variables, as the paper is focused on beams, the only displacement with incremental perturbation is the axial $u_1(x^1) = u_1^{\mathrm{F}}(x^1) + u_1^{\mathrm{I}}(x^1)$; other independent variables will only exhibit membrane behavior. With the new definition, the variation of the potential energy is defined as:

$$\mathcal{G}(\Delta\Phi) = \mathcal{G}(\Phi^{\mathrm{F}} + \Phi^{\mathrm{I}}) - \mathcal{G}(\Phi^{\mathrm{F}}) \geq 0 \qquad (15)$$

This expression is expanded in a Taylor's expansion, as in:

$$\mathcal{G}(\Delta\Phi) = \mathcal{G}(\Phi, \delta\Phi) + \frac{1}{2!}\mathcal{G}(\Phi, \delta\Phi, \delta^2\Phi) + \frac{1}{3!}\mathcal{G}(\Phi, \delta\Phi, \delta^2\Phi, \delta^3\Phi) + \cdots \qquad (16)$$

Based on Brush and Almroth [23], the above expansion is reduced into the second variational term by the minimum energy theorem, which its second term is positively defined for stability phenomenon. Therefore, the critical buckling load is obtained when the variation of the potential energy is zero, leading to:

$$\delta[\mathcal{G}(\Phi, \delta\Phi, \delta^2\Phi)] = 0 \qquad (17)$$

Replacing our fundamental-incremental scheme in the above expression, the following is derived:

$$\delta\left[\mathcal{G}\left(\Phi^{\mathrm{I}}, \delta\Phi^{\mathrm{I}}, \delta^2\Phi^{\mathrm{I}}\right)\right] + \delta\left[\mathcal{G}\left(\Phi^{\mathrm{FI}}, \delta\Phi^{\mathrm{FI}}, \delta^2\Phi^{\mathrm{FI}}\right)\right] = 0 \qquad (18)$$

The expression in equation (18) states that the potential energy can be divided into the incremental displacement and stress resultants, and into the fundamental-incremental terms, which define the buckling load and its eigenvalues.

3. Results

A Finite Element procedure is developed using a 4-element mesh with $\boldsymbol{p} = \mathbf{4}$ interpolation functions for primary variables to avoid shear locking problems. The adopted basis functions are $\boldsymbol{C}^{\mathbf{0}}$ interpolant polynomials of Gauss–Lobatto-Legendre (GLL) quadrature points [24], which are suitable for high-order expansions. Explicitly, the one-dimensional basis functions of the order $\boldsymbol{p} = \boldsymbol{m} - \mathbf{1}$ are expressed using the p-order Legendre polynomial $\boldsymbol{P}_{\boldsymbol{m}-\mathbf{1}}$, as shown

$$\phi^{(j)}(\xi) = \frac{(1-\xi^2)P'_{m-1}(\xi)}{m(m-1)P_{m-1}(\xi_j)(\xi - \xi_j)} \qquad (19)$$

For the validity of the proposal, nondimensional buckling stresses for various length to height ratios of simply-supported micropolar beams are compared against Hasanyan and Waas [25]. Additional results are presented in the last section using experimental material parameters obtained from [26] for micropolar and functionally graded beams.

3.1. Validation of the proposed model

The proposal is compared against the results of Hasanyan and Waas [25] for homogeneous beams. Nondimensional parameters are used according to the following equations

$$\hat{\kappa} = \kappa/\bar{\mu}\,, \qquad \hat{\lambda} = \lambda/\bar{\mu}\,, \qquad \hat{\beta}_3 = \beta_3\eta^2/\bar{\mu},$$
$$\hat{P}_{cr} = P_{cr}/(2A\bar{\mu})\,, \qquad \hat{s} = 2L/h\pi, \qquad \eta = \pi/L, \tag{20}$$

where A is the transversal area of the beam, P_{cr} is the buckling load of the beam and $\hat{s}$ is a slenderness ratio.

In figure 1, the normalized buckling load $\hat{P}_{cr}$ is compared for different values of $\hat{\kappa}$, with $\hat{\lambda} = 2$ and $\hat{\beta}_3 = 0$. The influence of the micropolar parameter κ in the buckling behavior of beams is shown. As the value of these parameters increases, the critical load is higher, which means that the stiffness of the beam increases. On the other hand, while the length to thickness ratio increases, results show convergence. Hence, as the slenderness of the beam grows, the influence of micropolar couple modulus on the buckling load decreases.

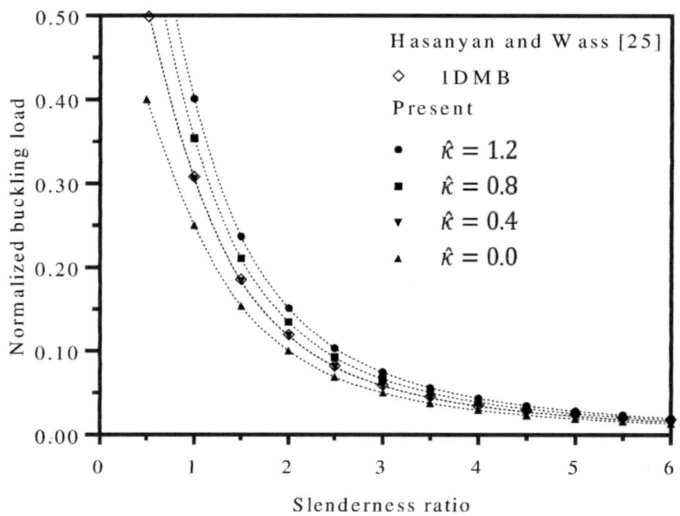

Figure 1. Normalized buckling load with $k \neq 0$

In figure 2, the normalized buckling load $\hat{P}_{cr}$ is analyzed for several values of $\hat{\beta}_3$, taking $\hat{\lambda} = 2$ and $\hat{\kappa} = 0$. The influence of $\hat{\beta}_3$ on the buckling load is evaluated as it is related to the curvature strain tensor on the micropolar continuum. The change of β_3 enhances the stiffness of the beam. Higher values of gamma turn into higher influence of these rotations on the overall stiffness.

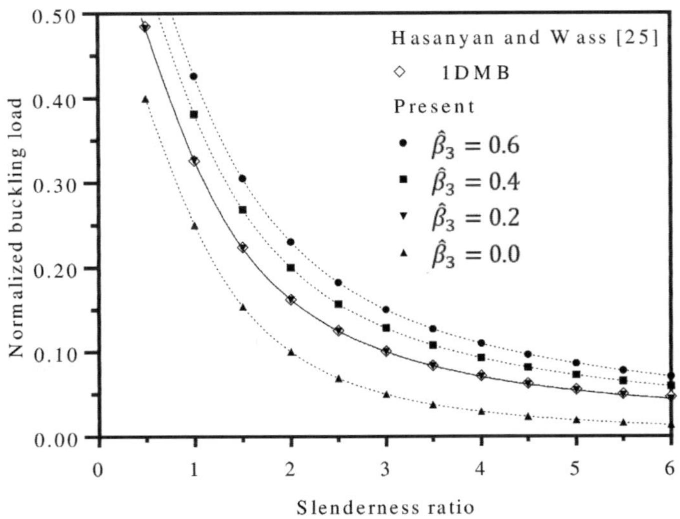

Figure 2. Normalized buckling load with $\beta_3 \neq 0$

3.2. Applications using experimental material properties in FGM micropolar beams

Additional results are further developed using experimental material parameters from the bibliography [26]. Simply-supported beams with different slenderness ratios are evaluated for various power-law exponents. The beam is composed of aluminum at the top with the following material parameters: $E_t = 70.85$ GPa, $\nu_t = 0.33$, $\kappa_t = 1.3155 \times 10^{-5}$ GPa, $\beta_{3_t} = 0.59664$ kN, $\beta_{2_t} = 0.1585$ kN ; and by epoxy at the bottom: $E_b = 5.31$ GPa, $\nu_b = 0.4$, $\kappa_b = 1.3234 \times 10^{-4}$ GPa , $\beta_{3_b} = 3.3349$ kN, $\beta_{2_b} = 0.1028$ kN. The poison ratios were used to calculate the Lamé coefficients for the two materials and then it was assumed an only varying Young modulus. The results of the analysis are shown in figure 3. The buckling load is normalized using $\bar{\mu}_t$ as in the previous examples for different values of the power-law index.

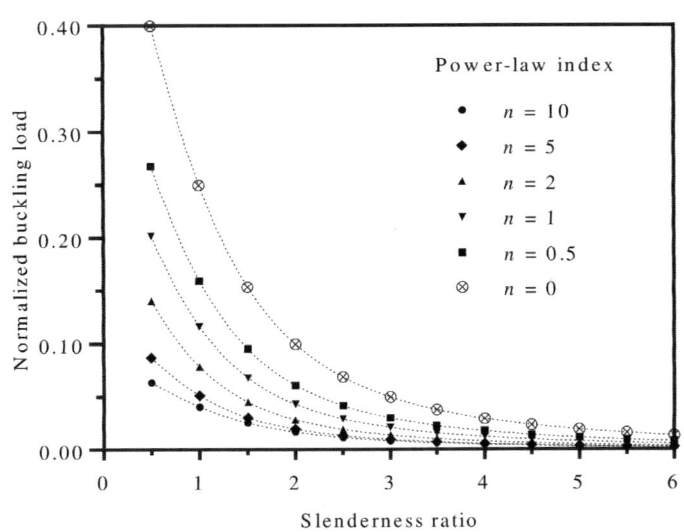

Figure 3. Normalized buckling load for various power-law exponents.

4. Conclusions

The present research has developed a finite element model for an improved first-order micropolar beam using the Eringen's kinematics tensors linearized by a Taylor expansion. High order terms where neglected. Furthermore, Trefftz's criterion has been used to evaluate the buckling phenomenon. First, the influence of the micropolar parameters on the buckling load was analyzed for different length to thickness ratios. Results were verified against the literature. It was shown that the couple modulus κ has a directly proportional relationship to the buckling load. However, as the length to thickness relation increases, its impact is reduced leading to a convergence of the results. Another parameter with a similar behavior is the micropolar modulus β_3. Although its influence is like κ, it does not vanish as the slenderness ratio of the beam increases. Finally, the effect of the power-law index in FGM micropolar beams was also calculated using material parameters found in the literature. New results for size-dependent micropolar parameters were obtained showing that the power-law index reduces the buckling load in micropolar beams. In the following research, results concerning geometrical non-linear behavior are expected using the non-linear micropolar stretch and wryness tensors for the analysis of the post-buckling phenomena in beams under different boundary conditions.

References

[1] Cuenot S, Frétigny C, Demoustier-Champagne S and Nysten B 2004 Surface tension effect on the mechanical properties of nanomaterials measured by atomic force microscopy *Phys. Rev. B - Condens. Matter Mater. Phys.* **69** 1–5

[2] Ma Q and Clarke D R 1995 Size dependent hardness of silver single crystals *J. Mater. Res.* **10** 853–63

[3] Cosserat, E., Cosserat F 1909 *Théorie des corps déformables* (Paris: A. Hermann et fils)

[4] Romanoff J, Jelovica J, Reddy J N and Remes H 2020 Post-buckling of web-core sandwich plates based on classical continuum mechanics: success and needs for non-classical formulations *Meccanica* **6**

[5] Nampally P, Karttunen A T and Reddy J N 2019 Nonlinear finite element analysis of lattice core sandwich beams *Eur. J. Mech. A/Solids* **74** 431–9

[6] Manzari M T 2004 Application of micropolar plasticity to post failure analysis in geomechanics *Int. J. Numer. Anal. Methods Geomech.* **28** 1011–32

[7] Mašín D, Tamagnini C, Viggiani G and Costanzo D 2006 Directional response of a reconstituted fine-grained soil - Part II : Performance of different constitutive models *Int. J. Numer. Anal. Methods Geomech.* **30** 1303–36

[8] Ansari R, Norouzzadeh A, Shakouri A H, Bazdid-Vahdati M and Rouhi H 2018 Finite element analysis of vibrating micro-beams and -plates using a three-dimensional micropolar element *Thin-Walled Struct.* **124** 489–500

[9] Penta F 2020 Buckling analysis of periodic Vierendeel beams by a micro-polar homogenized model *Acta Mech.* **231** 2399–424

[10] Carrera E and Zozulya V V. 2019 Carrera unified formulation (CUF) for the micropolar beams: Analytical solutions *Mech. Adv. Mater. Struct.* **0** 1–25

[11] Ramezani S, Naghdabadi R and Sohrabpour S 2009 Analysis of micropolar elastic beams *Eur. J. Mech. A/Solids* **28** 202–8

[12] Betancourt K N, Soncco K and Arciniega R A 2020 Bending Analysis of Micropolar Beams *LACCEI 2020 Proceedings* (accepted for publication) pp 29–31

[13] Hassanpour S and Heppler G R 2016 Comprehensive and easy-to-use torsion and bending theories for micropolar beams *Int. J. Mech. Sci.* **114** 71–87

[14] Hirano T, Yamada T, Teraki J, Niino M and Kumakawa A 1988 A study on a functionally gradient material design system for a thrust chamber *Proc. 16th Int. Symp. Sp. Technol. Sci.*

[15] Eringen A and Wegner J 2003 *Nonlocal Continuum Field Theories* vol 56

[16] Eremeyev, V.A.,Lebedev, L.P., Altenbach H 2013 *Foundations of Micropolar Mechanics* (Springer-Verlag Berlin Heidelberg)

[17] Cowin S C 1970 An incorrect inequality in micropolar elasticity theory *Zeitschrift für Angew. Math. und Phys. ZAMP* **21** 494–7

IOP Conf. Series: Materials Science and Engineering **999** (2020) 012017 doi:10.1088/1757-899X/999/1/012017

[18] Arciniega R A and Reddy J N 2007 Large deformation analysis of functionally graded shells *Int. J. Solids Struct.* **44** 2036–52

[19] Hasanyan A D and Waas A M 2015 On the buckling of a two-dimensional micropolar strip *J. Appl. Mech. Trans. ASME* **82**

[20] Pietraszkiewicz W and Eremeyev V A 2009 On natural strain measures of the non-linear micropolar continuum *Int. J. Solids Struct.* **46** 774–87

[21] Arciniega R A, Gonçalves P B and Reddy J N 2004 Buckling and postbuckling analysis of laminated cylindrical shells using the third-order shear deformation theory *Int. J. Struct. Stab. Dyn.* **4** 293–312

[22] Ayala S, Vallejos A and Arciniega R A 2020 Buckling Analysis of Functionally Graded Timoshenko Beams *LACCEI 2020 Proceedings* (To be submitted)

[23] Brush D O and Almroth B O 1975 *Buckling of Bars, Plates and Shells* (Mcgraw-hill)

[24] Karniadakis, G. E.; Sherwin S 2005 *Spectral/hp Element Methods for CFD.*

[25] Hasanyan A D and Waas A M 2015 On the Buckling of a Two-Dimensional Micropolar Strip *J. Appl. Mech.* **82**

[26] Abadikhah H and Folkow P D 2017 A systematic approach to derive dynamic equations for homogeneous and functionally graded micropolar plates *Procedia Eng.* **199** 1429–34

AUTHOR INDEX